URBAN FORESTS
Ecosystem Services and Management

URBAN FORESTS
Ecosystem Services and Management

Edited by

J. Blum, PhD

AAP APPLE ACADEMIC PRESS

Apple Academic Press Inc.
3333 Mistwell Crescent
Oakville, ON L6L 0A2
Canada

Apple Academic Press Inc.
9 Spinnaker Way
Waretown, NJ 08758
USA

Exclusive worldwide distribution by CRC Press, a member of Taylor & Francis Group

No claim to original U.S. Government works

Printed in the United States of America on acid-free paper

International Standard Book Number-13: 978-1-77188-425-9 (Hardcover)

International Standard Book Number-13: 978-1-77188-426-6 (eBook)

Library and Archives Canada Cataloguing in Publication

Urban forests : ecosystem services and management/edited by J. Blum, PhD.

Includes bibliographical references and index.
Issued also in print and electronic formats.
ISBN 978-1-77188-425-9 (hardcover).--ISBN 978-1-77188-426-6 (pdf)

1. Urban forestry. 2. Trees in cities. 3. Urban ecology (Biology). 4. Ecosystem services.
5. Ecosystem management. I. Blum, Janaki, editor

SB436 U73 2016 635.9'77 C2016-903215-9 C2016-903216-7

Library of Congress Cataloging-in-Publication Data

Names: Blum, Janaki, editor.
Title: Urban forests : ecosystem services and management / editor: J. Blum, PhD.
Description: Waretown, NJ : Apple Academic Press, [2016] | Includes bibliographical references and index.
Identifiers: LCCN 2016020837 (print) | LCCN 2016023075 (ebook) | ISBN 9781771884259 (hardcover : alk. paper) | ISBN 9781771884266 ()
Subjects: LCSH: Urban forestry. | Trees in cities--Environmental aspects. | Tree planting.
Classification: LCC SB436 .U722 2016 (print) | LCC SB436 (ebook) | DDC 634.909173/2--dc23
LC record available at https://lccn.loc.gov/2016020837

Apple Academic Press also publishes its books in a variety of electronic formats. Some content that appears in print may not be available in electronic format. For information about Apple Academic Press products, visit our website at **www.appleacademicpress.com** and the CRC Press website at **www.crcpress.com**

ABOUT THE EDITOR

J. BLUM, PhD

J. Blum, PhD, has a background in biology as well as urban and environmental policy and planning. Her experience in education, field work, research, evaluation, and project coordination is diverse, and includes biotechnology, food systems and urban agriculture, affordable housing policy, and small-scale economic development. She has studied or worked in educational, non-profit and for-profit environments in the USA, Europe, and Asia.

CONTENTS

ACKNOWLEDGMENT AND HOW TO CITE

The editor and publisher thank each of the authors who contributed to this book. The chapters in this book were previously published elsewhere. To cite the work contained in this book and to view the individual permissions, please refer to the citation at the beginning of each chapter. Each chapter was read individually and carefully selected by the editor; the result is a book that provides a multi-perspective look at research into many elements of urban forests. The chapters included are organized into four sections as follows:

The first section focuses on the various benefits of urban forests.

- Chapter 1 models urban air quality and climate regulation to examine the complex relationship between municipal forests, ecosystem services, and urban space, emphasizing the need for integrated green strategies.
- Chapter 2 demonstrates that well managed urban tree cover can contribute to urban sustainability through fossil fuel replacement, as well through carbon storage.
- Chapter 3 studies connections between tree cover and human health to suggest cost-effective strategies to improve well-being in certain urban populations.

The second section explores the issues surrounding urban tree planting.

- Chapters 4 and 5 explore the multifaceted response of urban residents to trees and tree planting, indicating a need for more citizen involvement in urban greening campaigns.

The third section centers on community-based stewardship of urban forests.

- Chapter 6 discusses a tool for communities could employ to gather data for the effective care of local trees.

The final section turns to enhancing knowledge of urban forestry.

- Chapter 7 analyzes urban tree dynamics via historical photography to evaluate conditions for preserving trees with high ecosystem value.
- Chapter 8 examines biological, social, and urban design factors that influence the survival of young urban trees in order to advance management practices.
- Chapter 9 scrutinizes urban forests patterns to find that socioeconomic factors strongly influence variations in urban tree cover diversity.
- Chapter 10 focuses on the effects of input management on new urban forests over time, suggesting that biological diversity increases as conditions favorable to invasive species recede.

LIST OF CONTRIBUTORS

Meghan L. Avolio
Department of Biology, University of Utah, Salt Lake City, UT, USA

Francesc Baró
Institute of Environmental Science and Technology (ICTA), Autonomous University of Barcelona (UAB)

Michael Battaglia
Michigan Technological University Research Institue

Marc G. Berman
Department of Psychology, The University of Chicago, Chicago, IL, USA; Grossman Institute for Neuroscience, Quantitative Biology, and Human Behavior, University of Chicago

Jessie Braden
City of New York, Department of Parks & Recreation, Central Forestry & Horticulture

Geoffrey L. Buckley
Ohio University

Lindsay K. Campbell
USDA Forest Service, Northern Research Station

Lydia Chaparro
Ecologistas en Acción

Lorraine Weller Clarke
Department of Botany and Plant Sciences, University of California, Riverside, Riverside, CA, USA

Daniel F. Díaz-Porras
Department of Animal and Plant Sciences, University of Sheffield, Sheffield, U.K; Escuela de Ciencias, Universidad Autónoma 'Benito Juárez' de Oaxaca, Oaxaca, Mexico

Karl L. Evans
Department of Animal and Plant Sciences, University of Sheffield, Sheffield, U.K

Jill L. Edmondson
Department of Animal and Plant Sciences, University of Sheffield, Sheffield, UK

Nancy Falxa-Raymond
Department of Ecology, Evolution, and Environmental Biology, Columbia University

Michael Feller
New York City Department of Parks & Recreation, Natural Resources Group

Alexander Felson
Yale University School of Architecture, School of Forestry & Environmental Studies

Burnell C. Fischer
The Vincent and Elinor Ostrom Workshop in Political Theory and Policy Analysis, and Center for the Study of Institutions, Population and Environmental Change, Indiana University, Bloomington

Michael Galvin
SavATree

Kevin J. Gaston
Environment and Sustainability Institute, University of Exeter, Penryn, Cornwall, UK

Thomas W. Gillespie
Department of Geography, University of California, Los Angeles, Los Angeles, CA, USA

Erik Gómez-Baggethun
Institute of Environmental Science and Technology (ICTA), Autonomous University of Barcelona (UAB)

Jennifer Greenfeld
Central Forestry & Horticulture, City of New York Department of Parks and Recreation

Morgan Grove
USDA Forest Service

Peter Gozdyra
Institute for Clinical Evaluative Sciences, Toronto, ON, Canada

G. Darrel Jenerette
Department of Botany and Plant Sciences, University of California, Riverside, Riverside, CA, USA

Omid Kardan
Department of Psychology, The University of Chicago, Chicago, IL, USA

Richard Karty
The New School, Tishman Environment and Design Center

Kristen L. King
Forestry, Horticulture, and Natural Resources, City of New York Department of Parks and Recreation

Jonathan R. Leake
Department of Animal and Plant Sciences, University of Sheffield, Sheffield, UK

Johannes Langemeyer
Institute of Environmental Science and Technology (ICTA), Autonomous University of Barcelona (UAB)

Jacqueline W.T. Lu
Forestry, Horticulture and Natural Resources, City of New York Department of Parks and Recreation

Heather R. McCarthy
Department of Microbiology and Plant Biology, University of Oklahoma, Norman, OK, USA

Nicola McHugh
Department of Animal and Plant Sciences, University of Sheffield, Sheffield, UK

P. Timon McPhearson
The New School, Tishman Environment and Design Center

Bratislav Misic
Indiana University, Bloomington, IN, USA

Faisal Moola
The David Suzuki Foundation, Toronto, ON, Canada

David J. Nowak
USDA Forest Service, SUNY-ESF

Odhran S. O'Sullivan
Department of Animal and Plant Sciences, University of Sheffield, Sheffield, UK

Lyle J. Palmer
Translational Health Science, The University of Adelaide, Adelaide, SA, Australia

Matthew I. Palmer
Columbia University, Department of Ecology, Evolution and Environmental Biology

Diane E. Pataki
Department of Biology, University of Utah, Salt Lake City, UT, USA

Stephanie Pincetl
Institute of Environment and Sustainability, University of California, Los Angeles, Los Angeles, CA, USA

Tomáš Paus
Rotman Research Institute, University of Toronto, Toronto, ON, Canada

Ruth A. Rae
City of New York, Department of Parks & Recreation, Central Forestry & Horticulture Division

Gabriel Simon
City of New York, Department of Parks & Recreation, Central Forestry & Horticulture

Erika S. Svendsen
USDA Forest Service, Northern Research Station

Jaume Terradas
Centre for Ecological Research and Forestry Applications (CREAF), Autonomous University of Barcelona (UAB)

Jessica M. Vogt
Center for the Study of Institutions, Population and Environmental Change, Indiana University, Bloomington

Tim Wenskus
New York City Department of Parks & Recreation, Natural Resources Group

INTRODUCTION

The term "urban forests" generally refers to all forms of vegetation that grow naturally, or through human activity, in and around densely settled human habitats. While this definition includes shrubs, grasses, and other plants, it is particularly directed at urban trees, both at individual and group levels, and in private as well as public settings. Much of the research on urban forests has been conducted in the United States of America and other highly urbanized areas. This book draws on examples from these countries to illustrate issues in urban forestry research and management. Associated quantitative and qualitative analytical tools are found throughout the volume, with their advantages or drawbacks discussed in the relevant chapters. It is hoped that these studies might afford insights applicable to a more global context.

Urban trees have long been appreciated for their esthetic appeal, but are increasingly valued as a means of mitigating the considerable human and environmental health problems associated with rapidly urbanizing areas. The range of benefits or "ecosystem services" that trees bestow on city-dwellers includes: air purification, carbon sequestration, temperature regulation, noise reduction, storm water management, and recreational opportunities. Thus trees are a part of urban "green infrastructure." However, they can also deliver "disservices" such as releasing volatile organic compounds, triggering pollen allergies, and damaging infrastructure. A vital part of urban forest research lies in understanding and quantifying these services and relationships so as to improve them through appropriate management, which includes both planting and care of urban trees. The ability of urban forests to provide ecosystem services depends upon leaf surface area, most pronounced in larger tree species. However, programs to expand this "leaf canopy" have been constrained by shrinking municipal budgets, resulting in partial devolution of urban forest management onto local communities, non-profits, and business sectors. Techniques for minimizing costs, and for collective strategizing, are being developed.

Community education and stewardship appear to be the most promising strategies with respect to promoting healthy urban forests. Identifying trends in urbanization and the accompanying growth and decline of urban forests through time will also aid in furthering urban forest management and green infrastructure planning.

Part 1 discusses four examples of benefits that urban forests provide. Baró et al. (chapter 1) turn to the much emphasized role of urban forests in reducing air pollution levels and offsetting greenhouse gas (GHG) emissions in cities. They demonstrate the complex nature of urban forest ecosystem services and disservices through a software application that models urban air quality and climate regulation in a major European city. They conclude that to be effective, urban forests and other green infrastructures must be implemented at broader spatial scales, and in conjunction with other "greening" strategies. McHugh et al. (chapter 2) go beyond the wind shielding effects of trees to reduce heating energy needs. Considering GHG emissions inherent in expanding urban forests, they show that by partially replacing fossil fuel with local timber, regularly coppiced urban tree cover can contribute to both climate regulation and energy conservation and therefore to urban sustainability. Kardan et al (chapter 3) brings in the direct connection of urban forest ecosystem services to people by studying the correlation between tree cover and health perception in a large city neighborhood, finding that planting more street trees could potentially provide a cost-effective way to increase the well-being of poorer urban residents to much higher levels.

Part 2 then details and analyzes the issues involved in expanding urban tree cover. Rae et al. (chapter 4) and Battaglia et al. (chapter 5) both unearth a complexity in residents' response to trees and tree planting in two cities that calls for more citizen involvement in such campaigns.

Part 3 consequently examines the management of urban forests by the communities involved, including results and tools. Vogt and Fisher (chapter 6) propose a standardized protocol that community members could employ to gather data necessary to evaluate the survival and growth of recently-planted urban trees.

Part 4 gives a sampling of research that examines urban forests over time, from past to future, in order to further better management practices. Diaz-Porras et al. (chapter 7) use photographs to examine the historical expansion and recession of urban tree cover in the UK in order to determine conditions under which large trees with high ecosystem value are preserved. Lu et al. (chapter 8) gained an initial understanding of factors, including siting, species, and community involvement, which are vital to the success of young street trees in one city, and that might inform management practices in others. Avolio et al. (chapter 9) provides a brief summary of research into environmental and sociological drivers of urban forests, finding that socio-economic, rather than biophysical, factors better explain variations in tree cover diversity in urban contexts. Lastly, McPhearson et al. (chapter 10) describe a long-term research program focusing on the impact of plant, soil and management interactions on new urban forests over time, hypothesizing an increase in biological diversity concomitant with a decrease in invasive species.

This book contains a selection of chapters aimed to provide a better understanding of urban forests through consideration of major ecosystem services and management regimes, particularly through a community lens. Together these underscore some of the challenges researchers and practitioners face to ensure the viability and vitality of urban forests and consequent human wellbeing.

PART I

THE BENEFITS OF URBAN FORESTS

CHAPTER 1

Contribution of Ecosystem Services to Air Quality and Climate Change Mitigation Policies: The Case of Urban Forests in Barcelona, Spain

FRANCESC BARÓ, LYDIA CHAPARRO,
ERIK GÓMEZ-BAGGETHUN, JOHANNES LANGEMEYER,
DAVID J. NOWAK, AND JAUME TERRADAS

1.1 INTRODUCTION

Urban forests, encompassing all trees, shrubs, lawns, and other vegetation in cities, provide a variety of ecosystem services to city-dwellers, such as air purification, global climate regulation, urban temperature regulation, noise reduction, runoff mitigation, and recreational opportunities, as well as ecosystem disservices, such as air quality problems, allergies, and damages on infrastructure (Escobedo et al. 2011; Gómez-Baggethun and Barton 2013; Gómez-Baggethun et al. 2013). Specifically, a significant body of literature has stressed the contribution of urban forests in reducing air pollution levels and offsetting greenhouse gas (GHG) emissions in cities (e.g., Jo and McPherson 1995; Beckett et al. 1998; McPherson et al. 1998; Nowak and Crane 2002; Yang et al. 2005; Nowak et al. 2006; Paoletti 2009; Zhao et al. 2010).

Air quality in cities is a major concern of the European Union (EU). In the last two decades, various policy instruments have been implemented at the European level to improve air quality in urban areas, mostly by regulating anthropogenic emissions of air pollutants from specific sources and sectors. These include the Directive 2010/75/EU on industrial emissions, the "Euro standards" on road vehicle emissions and the Directive 94/63/EC on volatile organic compounds emissions from petrol storage and distribution, among others. Yet, the last annual report on air quality in Europe (EEA 2013) estimated that many urban inhabitants in the EU are still exposed to air pollutant concentrations above the EU's legally binding limits (mainly set in the Directive 2008/50/EC on ambient air quality and cleaner air for Europe). For example, the report noted that 22–33 % of the urban population within the EU was exposed to particulate matter (PM10) concentrations above the 24-h average limit value (50 $\mu g\ m^{-3}$) during the period 2009–2011. This estimation of exposure increases dramatically (85–88 %) if it takes as reference the maximum levels recommended by the World Health Organization (WHO), currently set at 20 $\mu g\ m^{-3}$ (annual mean).

As for climate change mitigation policy, the member states of the EU committed to reduce their GHG emissions by at least 20 % from 1990 levels before the end of 2020 (Climate and Energy Package, EC 2008). In an attempt to extent this commitment at the local level, the European Commission launched the "Covenant of Mayors" in 2008. This initiative involves local authorities, voluntarily committing themselves to implement more sustainable energy policies within their territories by reducing GHG emissions at the local level by at least 20 % until 2020. Such action by local authorities is deemed critical to meet global climate change mitigation targets because some 80 % of worldwide energy consumption and GHG emissions are associated with urban activities (Hoornweg et al. 2011).

The focus of urban policy-making to meet the EU targets for both air quality and climate change mitigation largely remains on technical measures such as the use of the best available technology, fuel composition requirements, energy efficiency, or renewable energy actions. The potential of urban green space in contributing to the compliance of these environmental targets is broadly neglected by urban policy-makers (Nowak 2006; Escobedo et al. 2011). Yet, a growing number of studies conclude that management of urban forests to enhance ecosystem services supply can be a cost-effective strategy to meet specific environmental standards or policy targets (e.g., Escobedo et al. 2008, 2010).

This research assesses ecosystem services and disservices provided by urban forests and it discusses their potential contribution in achieving air pollution regulation policy targets in cities. The objectives are twofold. First, we quantify in biophysical accounts and monetary values two ecosystem services ("air purification" and "global climate regulation") and one ecosystem disservice ("air pollution" associated with biogenic volatile organic compounds (BVOC) emissions) generated by the urban forests in Barcelona, Spain. Second, we evaluate the potential of these ecosystem services to the achievement of environmental policy targets based on their actual contribution relative to air pollution and GHG emissions levels at the city scale. Accounting also the disservice allows having a "net" estimate of this contribution, since BVOC emissions from urban forests can negatively impact air quality of cities (Nowak et al. 2000).

1.2 MATERIALS AND METHODS

1.2.1 CASE STUDY: BARCELONA CITY

We conducted our research within the administrative boundaries of the municipality of Barcelona, Spain (Fig. 1). With 1.62 million inhabitants in an area of 101.21 km^2 (Barcelona City Council Statistical Yearbook 2012), Barcelona is the second largest city in Spain and one of the most densely populated cities in Europe (16 016 inhabitants km^{-2}).

The total green space[1] within the municipality of Barcelona amounts to 28.93 km^2 representing 28.59 % of the municipal area and a ratio of 17.91 m2 per inhabitant (Barcelona City Council Statistical Yearbook 2012). Most of this green space, however, corresponds to the peri-urban forest of Collserola (protected as a natural park). The inner-city of Barcelona (excluding Collserola) embeds only 10.98 km^2 of green space (Barcelona City Council Statistical Yearbook 2012), which amounts to 10.85 % of the municipal area and a ratio of 6.80 m^2 of green space per inhabitant. This ratio is very low in contrast to other European cities—especially in

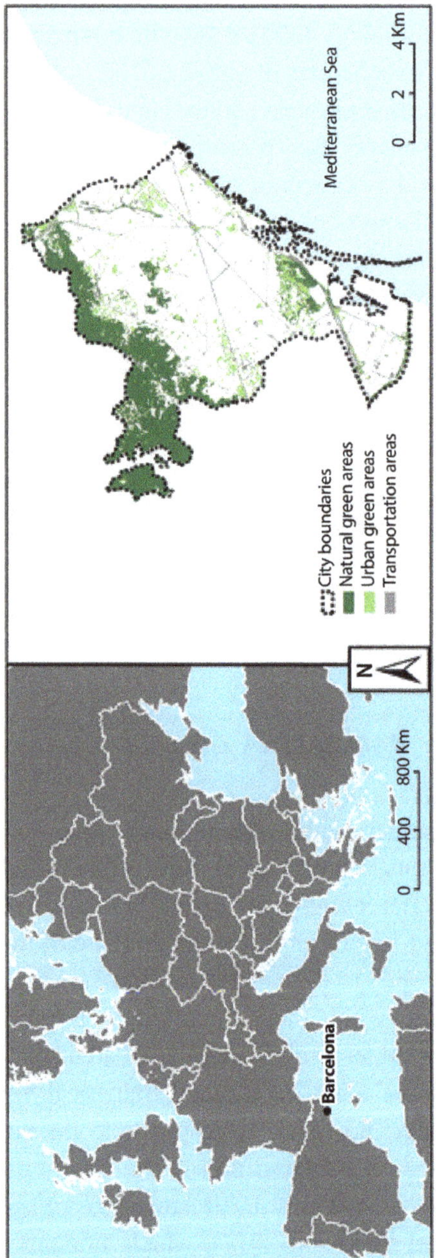

FIGURE 1: Location of Barcelona municipality and main green spaces. Source: Own elaboration based on Natural Earth datasets (www.naturalearthdata.com) and 3rd edition of the Ecological Map of Barcelona (Burriel et al. 2006)

northern countries—where green space amounts to up to 300 m^2 per inhabitant (Fuller and Gaston 2009). Nonetheless, these low levels of green space are partly counterbalanced by the high number of single street trees, accounting for 158 896 specimens in 2011, a ratio of 98.36 street trees per 1000 inhabitants. This ratio is relatively high compared to other urban areas in Europe, which mostly ranges between 50 and 80 street trees per 1000 inhabitants (Pauleit et al. 2002). Two species, *Platanus hispanica* (46 779 trees) and *Celtis australis* (19 426 trees), account for almost one-third of the street trees in Barcelona (Barcelona City Council Statistical Yearbook 2012). Thanks to recent research (e.g., Chaparro and Terradas 2009; Terradas et al. 2011), the role of urban forests in the provision of ecosystem services in Barcelona is starting to be acknowledged by the City Council as manifested, for example, in the *Barcelona Green Infrastructure and Biodiversity Plan 2020* (2013), a planning instrument that aims to aid the development of green infrastructure[2] (GI) strategies in the present decade.

As for many other large European cities (EEA 2013), air quality improvement stands as one of the major environmental policy challenges for Barcelona. In the last decade, the city has repeatedly exceeded the EU limit values for average annual concentrations of nitrogen dioxide (NO_2) and PM_{10} pollutants (40 $\mu g\ m^{-3}$ for both pollutants). The measures from the municipal monitoring stations during the period 2001–2011 show a steady trend for NO_2 values and a minor decrease for PM_{10} since 2006 (ASPB air quality report 2011). During the same period, ground-level ozone (O_3) levels have frequently exceeded the EU target value for human health (120 $\mu g\ ^{m-3}$ for a daily maximum 8-h mean period), but have never surpassed the number of allowed exceedances (25 days per year averaged over three years). Finally, carbon monoxide (CO) and sulfur dioxide (SO_2) concentrations have been historically very low in the city of Barcelona, never exceeding the EU limit values (125 $\mu g\ m^{-3}$ in one day for SO_2 and 10 mg m^{-3} for 8-h average for CO) (ASPB air quality report 2011). Figure 2 synthesizes the EU limit values for air quality and the maximum levels measured in Barcelona during 2011.

In 2008, Barcelona generated approximately 4.05 million metric tons of carbon dioxide equivalent (CO_2eq) emissions, mainly due to energy

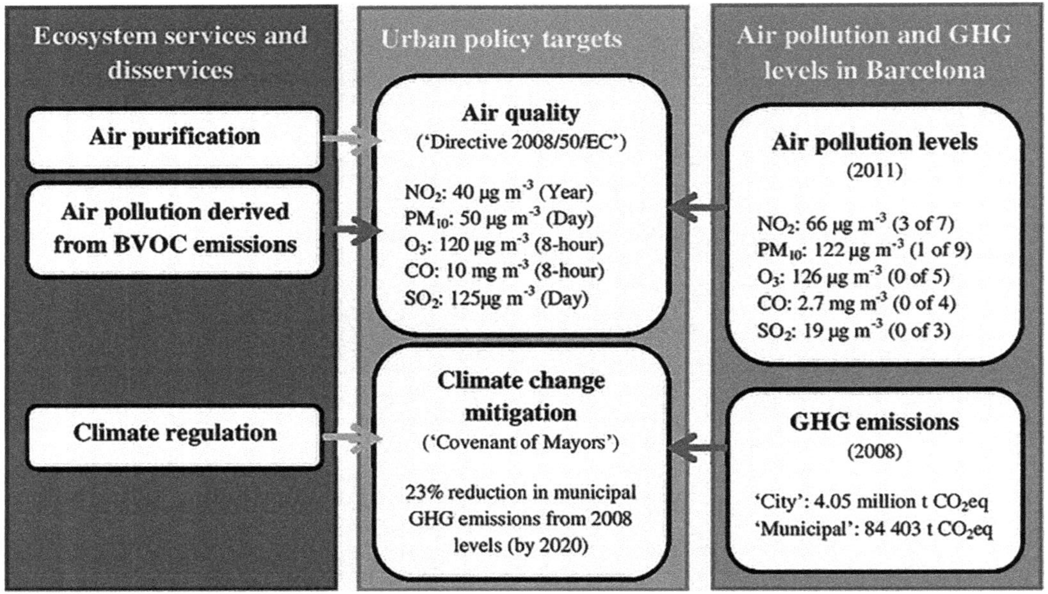

FIGURE 2: Framework for assessing links between ecosystem services and disservices, urban policy targets, and air pollution and GHG levels in Barcelona. Notes: air quality policy limits correspond to the most stringent EU values set for the protection of human health (in brackets the averaging period applicable for each limit). Some limits are subject to a specific number of allowed exceedances (e.g., PM_{10} limit can be exceeded 35 days per year at the most). See EEA (2013) for more details. Air pollution levels in Barcelona show the highest concentration values among all the monitoring stations measuring the corresponding air pollutant during the year 2011 (in brackets the number of monitoring stations exceeding the air quality limit after considering the number of allowed exceedances). See ASPB air quality report (2011) for more details. Arrows represent the links between ecosystem services and disservices, air pollution and GHG levels and urban policy targets in Barcelona (red arrows represent a negative impact towards policy targets and green arrows a positive impact). Sources: Own elaboration based on EEA (2013), ASPB air quality report (2011) and PECQ (2011)

consumption in the transportation, industry, housing, and services sectors (PECQ 2011). Compared to other cities worldwide, the ratio of Barcelona (2.51 t CO_2eq per inhabitant) is one of lowest proportions (Dodman 2009; Kennedy et al. 2009). This same year, the City Council of Barcelona signed the "Covenant of Mayors," committing to reduce by 23 % GHG emissions only derived from services and activities directly managed by the City Council by 2020 (this so-called "municipal" GHG emissions include emissions from municipal buildings, street lighting, municipal vehicle fleet and waste collection, among others). In 2008 (baseline year for Barcelona), municipal CO_2eq emissions amounted to 84 403 t, a ratio of 0.052 t per inhabitant (PECQ 2011, see Fig. 2).

The Energy, Climate Change and Air Quality Plan of Barcelona (PECQ 2011) provides the framework policy for air quality regulation and climate change mitigation during the period 2011–2020. Like other policy instruments aimed at improving indicators of environmental quality, the PECQ does not consider the enhancement of green infrastructure as a potential strategy to meet the policy targets established for air pollution concentrations and GHG emissions, as it focuses mainly on measures to improve energy efficiency and other technical fixes.

1.2.2 SAMPLE DESIGN AND DATA COLLECTION

The i-Tree Eco model (formerly known as Urban Forests Effects—UFORE) (Nowak and Crane 2000) was used to quantify ecosystem services and disservices in Barcelona. The i-Tree Eco model has been used in more than 50 cities across the world, especially in the United States, to assess urban forest structure and ecosystem services (Nowak et al. 2008a).

I-Tree Eco protocols (Nowak and Crane 2000; Nowak et al. 2008a, b; i-Tree User's Manual 2008) were followed to collect field data on urban forest structure within the municipality of Barcelona. Field data were collected within 579 randomly located circular plots (each measuring 404 m^2; 11.34 m radius) distributed across the city and pre-stratified among eight land use classes based on the 3rd edition of the Ecological Map of Barcelona (Burriel et al. 2006, see Fig. 3). Plot centers were positioned

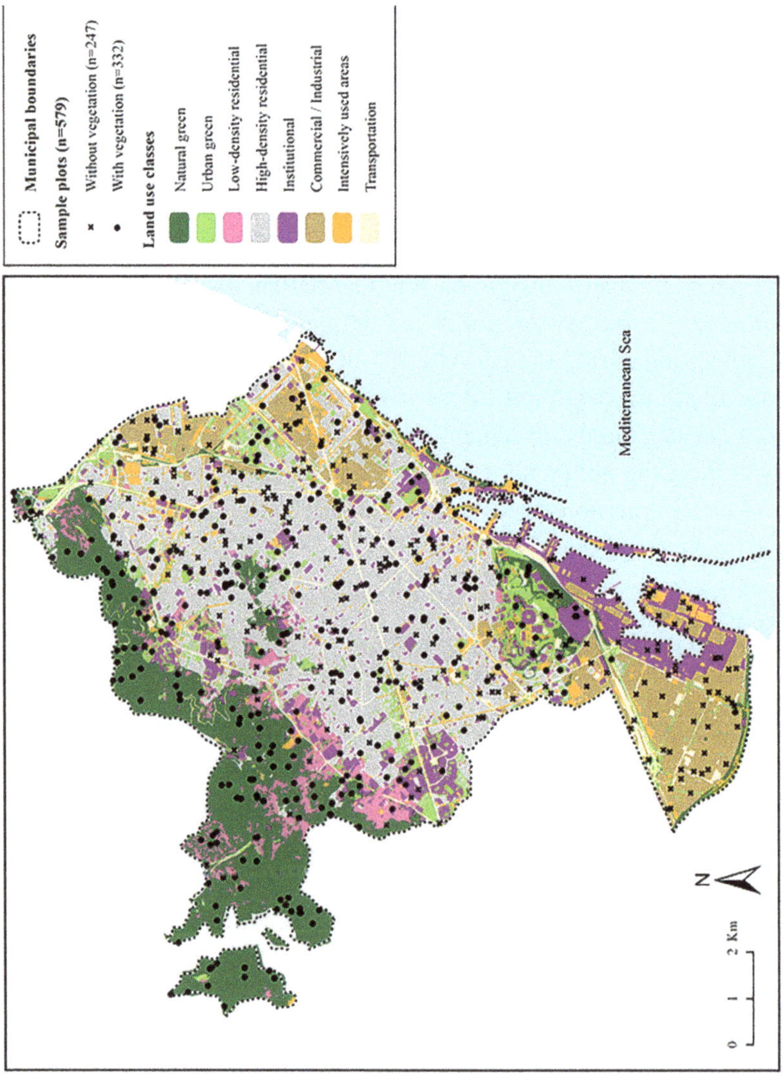

FIGURE 3: Land use classes and location of sample plots within the municipality of Barcelona. Source: Own elaboration based on the 3rd edition of the Ecological Map of Barcelona (Burriel et al. 2006)

from a random number generator of x and y coordinates for each land use class by means of a geographic information system (Miramon software, see Pons 2006). Prior to fieldwork, plots without vegetation cover were identified using 1:5000 digital aerial ortho-photographs from the Catalan Cartographic Institute (year 2004). Only the plots with vegetation cover (trees, shrubs or herbaceous flora) were then visited for field data collection (see Table 1 for sample data general figures).

Fieldwork was carried out from May to July 2009. Plots were located using a GPS device supported by high resolution maps containing the precise position of the plot center and its perimeter. Inaccessible plots (due to the steep slope, lack of permission to enter private areas, impenetrable vegetation, among others) were relocated in the closest accessible area with similar land use and vegetation characteristics. The general information collected from each visited plot included, among other parameters, date of visit, GPS coordinates, actual land use (and percent of land uses if the plot fell in more than one land use class), and percents of tree cover, shrub cover, plantable space, and ground cover. Main data on shrubs included the identification of species (genus at a minimum), average height, and percent area relative to total ground area. These data were collected for shrub masses (same species and height) and not at the individual level. Main data on trees included the identification of species, diameter at breast height (DBH), total height, height to crown base, crown width, percent of canopy missing (relative to crown volume), percent of impervious soil beneath canopy, percent of shrub cover beneath the canopy, and light exposure of the crown (see Nowak et al. 2008a for a complete list of data measures). Requirements of data inputs also include hourly air pollution concentrations and meteorological data (e.g., air temperature, solar radiation, and precipitation averages) for a complete year. The Public Health Agency of Barcelona (ASPB) provided concentration data for CO, SO_2, O_3, NO_2, and PM_{10} air pollutants from the 13 operational monitoring stations of the city during the year 2008. Meteorological data of Barcelona was directly retrieved from the US National Climatic Data Center (year 2008). Thus, the results from the evaluation of ecosystem services and disservices correspond to the year 2008.

Table 1. Sample data by land use stratification.

Land use class	Description[a]	Total area (ha)	Sample data				
			Sampled area (ha)	No. of plots	No. of plots with woody vegetation[b]	No. of trees	No. of shrub masses[c]
Urban green	Urban parks, lawns, allotment gardens, permanent crops, flowerbeds	806	2.02	50	50	544	89
Natural green	Woodland, scrubland, grassland, riparian vegetation, bare rock	2184	5.05	125	117	1844	329
Low-density residential	1–2 family dwellings (normally with private garden)	424	0.81	20	15	174	55
High-density residential	Multi-family dwellings with or without commercial areas	3666	8.24	204	102	531	79

Table 1. Continued.

Transportation	Parking lots, roads, rails and streets, stations	513	1.21	30	14	69	10
Institutional	Education, health, military, sport and other public facilities, cemeteries, port	776	1.58	39	3	21	0
Commercial/ industrial	Factories and other industrial areas, ware-houses, large shopping centers	1185	2.83	70	7	14	0
Intensively used areas	Pedestrian areas, vacant areas, areas in transfor-mation	567	1.66	41	24	148	8
Total		10 121	23.39	579	332	3345	570

[a] Based on land use subclasses from the 3rd edition of the Ecological Map of Barcelona (Burriel et al. 2006)
[b] Plots with woody vegetation account for those whether with shrubs or trees, or both
[c] Data on shrubs were collected for shrub masses (same species and height) and not at the individual level

1.2.3 QUANTIFICATION AND VALUATION OF ECOSYSTEM SERVICES AND DISSERVICES

Field data of urban forest structure, air pollution, and meteorological data were processed using i-Tree Eco software (www.itreetools.org) to quantify the ecosystem services of air purification and climate regulation, and the disservice air pollution derived from BVOC emissions in both biophysical and economic terms. Besides, the model also provided general results on the urban forest structure of Barcelona, including information on species composition, species origin and diversity, leaf area index (LAI), and leaf biomass. The analysis of the urban forest structure of Barcelona is beyond the scope of this paper; however, we refer to some relevant information in "Discussion" section.

The air purification service was quantified on the basis of field data, air pollution concentration, and meteorological data. Fundamentally, the i-Tree Eco model estimates dry deposition of air pollutants (i.e., pollution removal during non-precipitation periods), which takes place in urban trees and shrub masses. The (removed) pollutant flux (F; in g m^{-2} s^{-1}) is calculated as the product of deposition velocity (V d; in m s^{-1}) and the pollutant concentration (C; in g m^{-3}). Deposition velocity is a factor computed from various resistance components (for more details see Baldocchi et al. 1987; Nowak and Crane 2000; Nowak et al. 2006, 2008a). Monetary values of the ecosystem service air purification were estimated in i-Tree Eco from the median externality values for each pollutant established for the United States (Murray et al. 1994) and adjusted by the producer's price index for the year 2007 (U.S. Department of Labor). Externality values applied to the case study are: NO_2 = 9906 USD t^{-1}, PM_{10} = 6614 USD t^{-1}, SO_2 = 2425 USD t^{-1}, and CO = 1407 USD t−1. Externality values for O_3 are set to equal the value for NO_2.

The ecosystem service of climate regulation was calculated based on the modeling results of gross carbon sequestration, net carbon sequestration (i.e., estimated net carbon effect after accounting for decomposition emission of carbon from dead trees), and carbon storage. The i-Tree Eco model calculates the biomass for each measured tree using allometric equations from the literature. Biomass estimates are combined with base growth rates, based on length of growing season, tree condition, and tree competition, to derive annual biophysical accounts for carbon storage and carbon sequestration. Several assumptions and adjustments are considered in the modeling process (for more details, see Nowak and Crane 2000, 2002; Nowak et al. 2008a). To estimate the monetary value associated with urban tree carbon storage and sequestration, biophysical accounts were multiplied by 78.5 USD t^{-1} carbon based on the estimated social costs of carbon dioxide emissions in the US for the year 2010 (discount rate 3 %, EPA 2010). Additionally, we considered GHG emissions generated by the municipal vehicle fleet dedicated to green space management (862.50 t CO_2eq according to PECQ 2011) as a proxy of total GHG emissions directly attributable to green space maintenance. Hence, this measure was subtracted from total net carbon sequestration estimate provided by urban forests (after applying the conversion factor 1 g C = 3.67 g CO_2eq).

The emission of BVOCs from trees and other vegetation can contribute to the formation of ground-level O_3 and CO air pollutants (Kesselmeier and Staudt 1999), hence counteracting the air purification that vegetation delivers. BVOC emissions depend on factors such as tree species, leaf biomass, daylight, and air temperature (Nowak et al. 2008a). The i-Tree Eco model estimates the hourly emission of isoprene (C5H8), monoterpenes (C10 terpenoids), and other BVOCs by trees and shrubs species using protocols of the Biogenic Emissions Inventory System (BEIS; see Nowak et al. 2008a for further details). To estimate the amount of O_3 produced by BVOC emissions, the model applies incremental reactivity scales (g O_3 produced per g BVOC emitted) based on Carter (1994). CO formation from BVOC emissions is estimated for an average conversion factor of 10 % based on empirical evidence (Nowak et al. 2002a). However, due to the high degree of uncertainty in the approaches of estimating O_3 and CO formation derived from BVOC emissions, no estimates of the total amount of pollution formed by urban forests are given (neither monetary costs). Only index values can be calculated to compare the relative impact of the different species on O_3 and CO formation (Nowak et al. 2002a).

1.2.4 CONTRIBUTION OF URBAN FORESTS TO AIR QUALITY IMPROVEMENT AND CLIMATE CHANGE MITIGATION

The relative contribution of urban forests to air quality improvement and climate change mitigation in Barcelona for the year 2008 was determined based on data of air pollution levels and GHG emissions. We considered emissions generated within the municipal area (hereafter city-based pollution) and pollution not directly attributable to city-based emissions (hereafter background pollution) to determine air pollution levels in the city. We only accounted for PM_{10} and NO_2 levels since, as described above, these are the two air pollutants whose concentrations are frequently exceeding EU value limits in the city. Data for city-based pollution and background pollution were extracted from PECQ (2011) estimations. PECQ (2011) measures include aggregated and disaggregated city-based emissions from different sectors (road transport, residential and tertiary, industry and energy generation, and port activity), which in turn draws on a wide range of primary data sources (e.g., vehicle population, annual vehicle mileage, consumption of gas in households and businesses, etc.) and apply various quantitative methods (e.g., COPERT/CORINAIR model for road transport). Background pollution is measured from real pollutant concentration values recorded by the monitoring stations in the city and from one monitoring station located in the area of "Cap de Creus" (130 km north-east from Barcelona), hence not influenced by polluting activities within the city. According to PECQ (2011), the annual average concentration of NO_2 for the year 2008 in Barcelona was mainly determined by emissions from road traffic (65.6 %), while background pollution only accounted for 18.7 %. In contrast, the annual average of the PM10 concentration was primarily determined by background pollution (88.1 %).

The rate of GHG emissions was also extracted from PECQ (2011). Calculations are based on the various energy sources generating GHG emissions in the city (mainly electricity, natural gas and vehicle fuels).

Electricity-related GHG emissions are calculated based on the Catalan electricity mix.

1.3 RESULTS

1.3.1 AIR PURIFICATION

Total air purification is estimated at 305.6 t of removed pollutants year^{-1} with an economic value of 2.38 million USD year^{-1} (Fig. 4). PM10 removal is the highest among the five air pollutants analyzed (i.e., CO, NO2, PM_{10}, O_3, and SO_2), accounting for 54 % of the total biophysical value (166.0 t year−1) and 46 % of the total economic value (1.10 million USD year^{-1}). Pollution removal was lower for NO_2 and ground-level O_3 (54.6 t, 541 000 USD for NO_2; 72.6 t, 719 000 USD for O_3), and lowest for CO and SO_2 (5.6 t, 7880 USD for CO; 6.8 t, 16 000 USD for SO_2).

Average values for monthly removal of air pollution show a similar pattern across pollutants. January, November, and December were clearly the months where the uptake was lowest for all pollutants (percentages of uptake during the 3 months were 4.58 for CO, 8.45 for NO_2, 15.15 for PM_{10}, 2.69 for O_3, and 6.75 for SO_2). Spring and summer (from April to September) were the seasons with higher removal rates in average (percent of uptake during the 2 seasons was 60.96 for CO, 64.25 for NO_2, 54.43 for PM_{10}, 78.90 for O_3, and 70.46 for SO_2), although in some cases the highest monthly uptake rate corresponded to other periods (e.g., PM_{10} removal was highest in February, accounting for 10.69 % of total uptake). These patterns in uptake values are normally correlated with the seasonal variation in air pollutants concentrations and the biological cycle of trees (Nowak 1994; Yang et al. 2005). For instance, removal rates of ground-level O_3 are highest in summer, when concentrations are normally higher due to a more active process of photochemical reaction forming O_3 as a consequence of warmer temperatures and due to increased leaf surface area and gas exchange at the leaf surface.

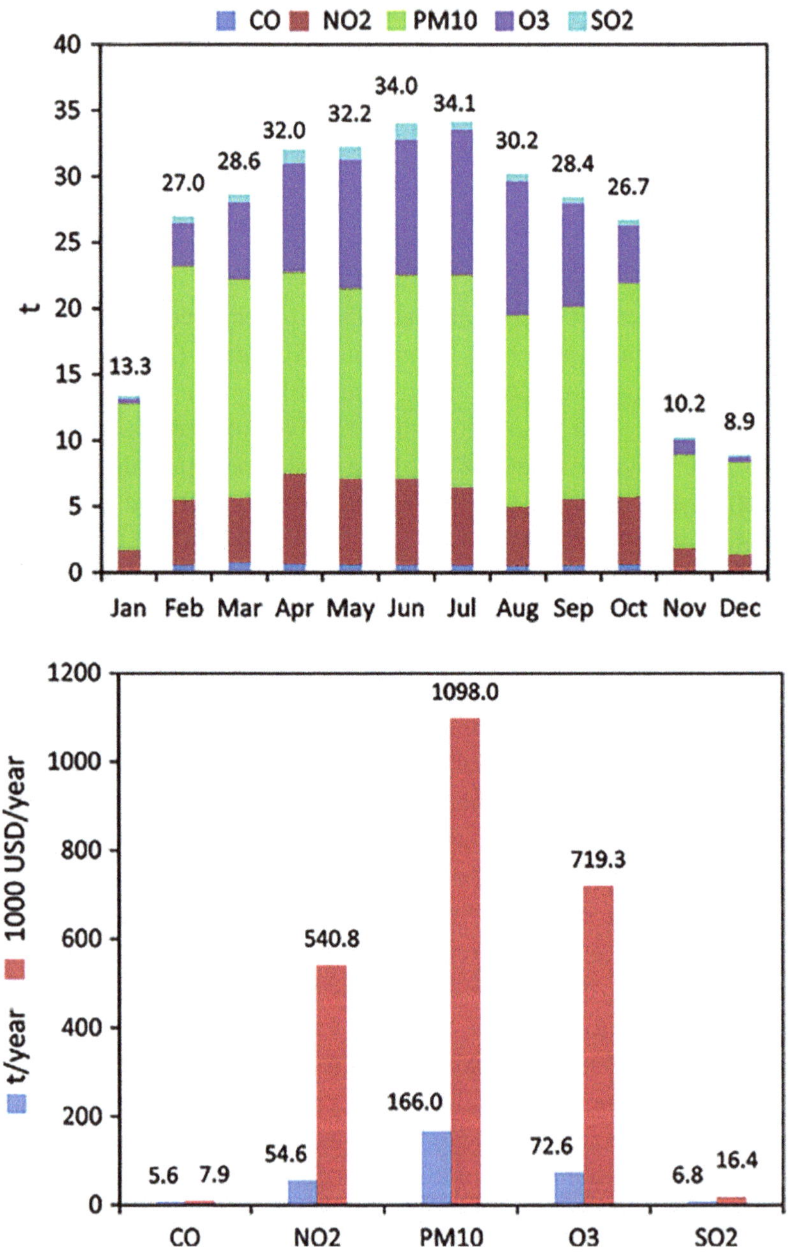

FIGURE 4: Monthly and annual air pollution removal by air pollutant (urban forests of the municipality of Barcelona, year 2008).

1.3.2 CLIMATE REGULATION

The total biophysical value of net carbon sequestration is estimated at 5187 t C year^{-1} (19 036 t CO_2eq year^{-1}) with an economic value of 407 000 USD year^{-1} (Table 2). This total net carbon sequestration is the only value including the effect of GHG emissions of green space maintenance, since disaggregate data by land use was not available. In absolute terms urban green, natural green, and high-density residential are the land use strata contributing the most to total net carbon sequestration (19, 39, and 24 %, respectively). However, considering the ratio net carbon sequestration per land use area, it is the urban green class that shows the highest values among these three land uses (1.24 t ha^{-1} urban green, 0.96 t ha^{-1} natural green, and 0.35 t ha^{-1} high-density residential). Surprisingly, the highest ratio among all land use classes is in the low-density residential stratum (1.33 t ha^{-1}).

1.3.3 AIR POLLUTION DUE TO BIOGENIC EMISSIONS

The total biophysical value of BVOC emissions is estimated at 183.98 t year^{-1} (Table 3). Similar to the case of carbon sequestration values, results for biogenic emissions show a major contribution of urban green, natural green, and high-density residential land use strata relative to the overall biophysical value for this ecosystem disservice (17.05, 47.46, and 15.32 %, respectively). Urban green, natural green, and low-density residential show to be the strata with the highest relative contribution to BVOC emissions in the city (39, 40, and 35 kg ha^{-1}, respectively) considering the ratio BVOC emissions per land use area. Besides, isoprene is clearly the main BVOC emitted (51.8 % of total emissions) in all land use classes (except for institutional), followed by other BVOCs (28.6 %) and monoterpenes (19.6 %).

Table 2. Carbon storage and annual carbon sequestration by land use class (urban forests of the municipality of Barcelona, year 2008).

Land use class	Biophysical values						Monetary values	
	Carbon storage		Gross carbon sequestration		Net carbon sequestration		Net carbon sequestration	
	t	SE	t $year^{-1}$	SE	t $year^{-1}$	SE	USD $year^{-1}$	SE
Urban green	26 876	4083	1088	109	1002	100	78 688	7839
Natural green	42 108	4115	2446	207	2099	181	164 804	14 224
Low-density residential	9764	2663	613	169	565	155	44 326	12 173
High-density residential	21 014	2940	1398	157	1282	149	100 630	11 66
Transportation	3876	1213	207	56	196	54	15 366	4250
Institutional	3452	2200	76	43	−64	109	−4995	8518
Commercial/ industrial	328	153	32	15	31	14	2409	1086
Intensively used areas	6020	1693	328	65	311	62	24 396	4844
Total	113 437	19 059	6187	819	5422 5187[a]	823	425 625 407 177[a]	64 595

[a] Net carbon sequestration values taking into account GHG emissions of green space maintenance
SE standard error

Table 3. Annual BVOC emissions by land use class (urban forests of the municipality of Barcelona, year 2008).

Land use class	Isoprene emissions (t year^{-1})	Monoterpenes emissions (t year^{-1})	Other BVOCs emissions (t year^{-1})	Total BVOC emissions (t year^{-1})
Urban green	16.78	4.94	9.65	31.36
Natural green	38.79	23.65	24.87	87.31
Low-density residential	8.81	1.93	4.06	14.81
High-density residential	17.09	3.20	7.89	28.18
Transportation	4.19	0.57	1.24	6.01
Institutional	0.91	1.18	2.69	4.78
Commercial/industrial	1.13	0.01	0.16	1.29
Intensively used areas	7.66	0.58	2.00	10.24
Total	95.36	36.07	52.56	183.98

1.3.4 ECOSYSTEM SERVICES CONTRIBUTION TO AIR QUALITY AND CLIMATE CHANGE MITIGATION

From total biophysical accounts for removal of PM_{10}, NO_2, and CO_2eq, we estimated the relative contribution of urban forests ecosystem services to air quality and climate change mitigation based on air pollution and GHG emissions levels in the city (Table 4). Our results suggest that the contribution of urban forests to climate change mitigation is very low, accounting for 0.47 % of the overall city-based GHG emissions. If we only account for GHG emissions derived from the sectors that are directly managed by the City Council (reference emissions to meet "Covenant of Mayors" 23 % reduction target and representing 2.10 % of the total emissions) the contribution of urban forest is still modest but yet substantial, accounting for 22.55 % of the emissions. Contributions of urban forests to air quality based only on city emissions differ notably depending on each air pollutant. While the overall contribution of urban forest to NO_2 removal is low relative to total emissions (0.52 %), its contribution to the removal of PM_{10} amounts to a significant 22.31 %. However, if we account for background pollution levels, the contribution of PM_{10} removal drops to 2.66 % of total PM_{10} pollution levels.

1.4 DISCUSSION

1.4.1 URBAN FORESTS POTENTIAL CONTRIBUTION TO MEET AIR QUALITY POLICY TARGETS

Urban forests effects on air quality are still a subject of intensive research. While positive effects of air purification delivered by vegetation have been estimated at the city scale in many urban areas (e.g., Nowak et al. 2006), pollution concentration can be increased at the site scale (e.g., street canyons) depending upon vegetation configuration, pollutant emissions, or meteorology, showing apparently divergent results on the effectiveness of using urban vegetation for reducing local air pollution hotspots (Pugh et al. 2012; Vos et al. 2013). Likewise, the ability of urban vegetation to

Table 4. Contribution of urban forests on air quality and climate change mitigation (year 2008).

Air pollutant	Removal biophysical value (t year^{-1})	Removal monetary value (USD year^{-1})	City-based emissions (t year^{-1})	Background pollution influence (%)	Ecosystem Service contribution (%)	
					City-based emissions	City-based emissions and background pollution
PM_{10}	166.01	1 097 964	743.77	88.10	22.32	2.66
NO_2	54.59	540 745	10 412.94	18.70	0.52	0.43
CO_2eq	19 036	407 177	4 053 766 84 403[a]	N/A	0.47 22.55[a]	N/A

[a] CO_2eq emissions from services and activities directly managed by the City Council ("Covenant of Mayors" policy target baseline emissions)

remove air pollutants significantly depends on many factors, such as tree health, soil moisture availability, leaf-period, LAI, meteorology, and pollution concentrations.

Our results show that the overall annual air purification rate by urban forests in Barcelona (9.3 g m^{-2} of canopy cover $year^{-1}$) is very similar to US cities like Columbus, Kansas City, or Portland (9.2 g m^{-2} $year^{-1}$), although the PM10 removal rate (5.1 g m^{-2} $year^{-1}$) is significantly higher than for these cities (between 3.1 and 3.4 g m^{-2}) and closer to cities like Salt Lake City (5.2 g m^{-2}), Philadelphia (5.5 g m^{-2}), or San Diego (5.6 g m^{-2}) (Nowak et al. 2006). The higher removal rates for PM_{10}, NO_2, and O3 compared to CO and SO_2 should be mainly attributable to the almost linear relationship between pollution removal and ambient pollution concentrations considered in the model (pollutant flux equation as F = V d × C). However, very high pollutant concentrations could severely damage vegetation or lead to stomatal closure, reducing air pollution removal ability (Robinson et al. 1998; Escobedo and Nowak 2009). Unfortunately, these environmental thresholds are not yet factored in the i-Tree Eco model.

Our findings also show that the NO_2 removal rate by urban forests in Barcelona has a meager impact relative to actual city-based emissions (less than 1 %). Therefore, the potential of urban forests to contribute to the compliance of the EU limit is expected to be very low. NO_2 concentrations in the city derive largely from road transport activity (65.6 % impact according to PECQ 2011). Hence, actions focused on reduction of road traffic, technological change toward less-polluting fuels and the promotion of public transport or cycling utilities are expected to contribute more efficiently to meet policy targets. These actions can also lead to reduction in O_3 concentrations, as NO_2 is a precursor chemical to O_3 formation. PM_{10} removal rate from urban forests is notably higher than NO_2 rate, whereas city-based emissions of PM_{10} are notably lower, resulting in a substantial impact at the city scale (22.3 % of total city-based emissions). However, the background pollution effect (accounting for 88.1 % of the average annual PM_{10} concentration according to PECQ estimations) drastically reduces the actual impact of the urban forests service (2.7 % of total PM_{10} pollution levels). Yet, we claim that there are still important reasons for which this ecosystem service should be accounted for in local policy decision-making. First, air pollution from particulate matter is a major health problem in Barcelona metropolitan area and recent research suggests that even moderate improvements in air quality are expected to report significant health benefits, together with related economic savings (Pérez et al. 2009). Second, the major role of PM_{10} background pollution in Barcelona air quality might compromise the effectiveness of municipal policies solely based on city emissions abatement. This fact also suggests that measures focused on air quality regulation should be implemented at broader spatial scales, particularly at the metropolitan level. To this end, strong coordination policies between municipal and regional authorities dealing with environmental quality and urban planning are fundamental. Third, the implementation of green infrastructure-based strategies to foster air purification (and other ecosystem services) is a realistic policy option considering the current urban context of Barcelona. I-Tree Eco results show that approximately 3.6 % of the municipality area (364 ha) can be considered as available land for planting. As a complementary alternative, green roofs and walls, yet to be extensively developed in Barcelona, could be particularly appropriate in high-density neighborhoods where ground for planting is extremely scarce. Several studies have quantified the potential of green roofs for air purification in cities at the street canyon (Baik et al. 2012), neighborhood (Currie and Bass 2008), and municipality (Yang et al. 2008) scales, besides their potential to provide many other services and benefits, such as runoff mitigation, noise reduction, or urban cooling (Oberndorfer et al. 2007; Rowe 2011). However, the technical and economic feasibility of green roofs expansion, together with possible trade-offs concerning their maintenance such as water demand, should previously be assessed in Barcelona, especially for existing buildings.

Proper management of existing green space can also contribute to air quality improvement. Yang et al. (2005) lists several factors to consider in strategies for air quality improvement based on green infrastructure, including selection of species (e.g., evergreen versus deciduous trees, dimension, growth rate, leaf characteristics, or air pollution tolerance) and management practices (e.g., intensity of pruning). Previous studies in cities with high levels of air pollution (e.g., Nowak et al. 2006; Escobedo and Nowak 2009) suggest that meteorological conditions, mixing-layer height (the atmospheric layer which determines the volume available for the dispersion of pollutants, see Seibert et al. 2000 for a complete definition), and vegetation characteristics (e.g., proportion of evergreen leaf area, in-leaf season, and LAI) are important factors defining urban forest effects on air quality. Further research is needed to advance our understanding of the role of morphology, function, and ecophysiology of vegetation in air purification (Manning 2008).

A further critical issue concerns the understanding of trade-offs with other ecosystem services or disservices. For example, urban parks are considered very relevant ecosystems for the provision of outdoor recreation and other cultural services in cities (Chiesura 2004). However, highly maintained parks might remove less air pollutants and CO_2 (due to emissions from maintenance activities, Nowak et al. 2002b) than natural areas that are not intensively managed, but which can be perceived as unpleasant or even dangerous, hence providing few cultural services (Lyytimäki and Sipilä 2009; Escobedo et al. 2011). Likewise, urban tree species with high potential for air purification can be highly invasive as well in certain cities (Escobedo et al. 2010). More generally, many specific environmental factors (e.g., soil condition, climate, water availability, or longevity of the species) should be considered in urban forest management to avoid conflicts with other municipal sustainability goals (Yang et al. 2005; Escobedo et al. 2011).

The i-Tree Eco model could not provide reliable results on O_3 and CO formation rates associated to the quantified BVOC emissions. However, as mentioned above, CO levels in Barcelona (2.7 mg m^3 for a daily 8-h average was the highest measure in 2011 according to ASPB air quality report 2011) have been historically far below the EU reference value (10 mg m^3 daily 8-h average). Thus, it is unlikely that urban forests may compromise in any significant form the compliance of air quality relative to CO target. In contrast, ground-level O_3 levels have surpassed the EU reference value (120 μg m^{-3} daily 8-h average) at some monitoring stations in the last decade, even if the allowed exceedences have never been reached. Although O_3 concentrations have remained steady in the last decade within the municipality of Barcelona, O_3 formation due to BVOC emissions might cause air quality problems in the long term, where BVOC emissions are expected to increase due to global warming (Peñuelas and Llusià 2003). Nevertheless, several studies point out that the selection of low BVOC-emitting tree species can contribute positively in O_3 concentrations in urban areas because BVOC emissions are temperature dependent and trees generally lower air temperatures (Taha 1996; Nowak et al. 2000; Paoletti 2009). Chaparro and Terradas (2009) identified some of the tree and shrub species in Barcelona emitting less BVOC per leaf biomass. These include genera such as Pyrus, Prunus, Ulmus, and Celtis.

1.4.2 URBAN FORESTS POTENTIAL CONTRIBUTION TO MEET CLIMATE CHANGE MITIGATION POLICY TARGETS

Some authors suggest that global climate regulation does not stand amongst the most relevant ecosystem services in the urban context because cities can benefit from carbon offsets performed by ecosystems located elsewhere (Bolund and Hunhammar 1999). However, other authors argue that urban forests can play an important role in mitigating the impacts of climate change if compared to other policies at the city level (McHale et al. 2007; Escobedo et al. 2010; Zhao et al. 2010; Liu and Li 2012).

The estimated net annual carbon sequestration per hectare of Barcelona (536 kg ha^{-1} $year^{-1}$) is very similar to cities such as Baltimore (520 kg ha^{-1} $year^{-1}$) or Syracuse (540 kg ha^{-1} $year^{-1}$) (Nowak and Crane 2002). It should be noted that an analysis of the overall contribution of urban green infrastructure to climate change mitigation should also account for the effects of vegetation on micro-climate regulation, which can indirectly avoid CO_2 emissions through energy saving in buildings for heating and cooling (Nowak and Crane 2002). Hence, our quantification likely underestimates the total contribution of urban forests to climate change mitigation. Analyzing the results by land use, urban green and natural green strata are relevant for the supply of climate regulation service due to the high vegetative cover compared to the other land use classes. High-density residential stratum also showed an important rate in net carbon sequestration, mainly attributable to its large total area (36 % of the municipality) and probably, to a lesser extent, to the high presence of street trees in these neighborhoods. Finally, the high ratio of net carbon sequestration per area observed in the low-density residential stratum could be attributed to the high presence of private gardens in these areas, together with low decomposition emissions due to healthier vegetation.

In line with the results obtained in other urban studies (Pataki et al. 2009; Liu and Li 2012), our findings show that direct net carbon sequestration in Barcelona makes a very modest contribution to climate change mitigation relative to total city-based annual GHG emissions (0.47 %). Nevertheless, if we only account for the GHG emissions from services and activities directly management by the City Council (baseline emissions for the 23 % reduction target from the "Covenant of Mayors"), the contribution of urban forest is notably higher (22.55 %). Similar green infrastructure-based strategies as specified for air quality improvement could also improve the contribution of urban forests to offset GHG emissions and meet the urban policy target of 23 % reduction until 2020.

1.4.3 LIMITATIONS AND CAVEATS

The main advantages of the i-Tree Eco model stem from the reliance on locally measured field data and standardized peer-reviewed procedures to measure urban forest regulating ecosystem services in cities (Nowak et al. 2008a). Favored by its status as an open access model, it has been widely applied across the world (e.g., Nowak and Crane 2002; Yang et al. 2005; Nowak et al. 2006; Currie and Bass 2008; Escobedo and Nowak 2009; Dobbs et al. 2011; Liu and Li 2012).

However, i-Tree Eco has some limitations that should be taken into account when analyzing its outcomes. First, the model is especially designed for US case studies and its application in other countries is subject to some restrictions, as stated in the user's manual. For instance, although the i-Tree Eco database has over 5000 species, it did not include some tree and shrub species sampled in Barcelona, which then needed to be added to the database. Likewise, monetary valuations of air purification and climate regulation services are based on the literature (see "Materials and Methods" section) which mainly apply to the US context and, hence, should be considered a rough estimation for Barcelona. However, these values are direct multiplier to the biophysical accounts, thus they can be easily adjusted to the case study context when data will be available. Another important limitation applying to i-Tree Eco and most dry deposition models is the level of uncertainty involved in the quantification of the air pollution removal rates due to the complexity of this process (Pataki et al. 2011). For instance, some sources of uncertainty include non-homogeneity in spatial distribution of air pollutants, particle re-suspension rates, transpiration rates, or soil moisture status (Manning 2008). Though the model outputs match well with field measured deposition velocities for urban forests, the model analyzes average effects across a city, not local variations in removal caused by local meteorological and pollution differences. However, these local fine-scale input data are often missing from urban areas and empirical data on the actual uptake of pollutants by urban vegetation are still limited (Pataki et al. 2011; Setälä et al. 2013), which makes a more accurate modeling of this ecosystem service unfeasible at the moment. For a sensitivity analysis of the i-Tree Eco deposition model see Hirabayashi et al. (2011). Estimation errors in climate regulation service values include the uncertainty from using biomass equations and conversion factors as well as measurement errors (Nowak et al. 2008a). For example, there are limited biomass equations for tropical tree species (e.g., palm trees), some of them present in Barcelona. Estimates of carbon sequestration and storage also include uncertainties from factors such as urban forests maintenance (e.g., intensity of pruning), tree decay, or restricted rooting volumes, which are not accounted for in the model's estimations (Nowak et al. 2008a; Pataki et al. 2011). BVOC emissions are estimated based on species factors and meteorological conditions (i.e., air temperature and daylight) but the uncertainty of the estimate is unknown. As mentioned in previous sections, O_3 and CO formation rates from BVOC emissions cannot be estimated with an acceptable level of reliability.

Therefore, the results presented in this paper should be considered as an approximate estimation rather than a precise quantification of the ecosystem services and disservices delivered by the urban forests of Barcelona. However, these estimates allow one to evaluate the contribution of urban forests in air quality and climate change mitigation in the city, and also to derive implications and recommendations for urban decision-making.

1.5 CONCLUSION

Regulating ecosystem services provided by urban forests have been widely analyzed in many cities across the world. However, the potential effectiveness of urban forests in air quality improvement and climate change mitigation is still object of debate, mainly due to the multiple factors and uncertainties involved in the actual delivery of these ecosystem services in cities, especially at the patch or site scale. Further, this potential is barely reflected in terms of its contribution to meet specific policy targets.

Our findings show that the contribution of urban forests regulating services to abate pollution is substantial in absolute terms (305.6 t of removed air pollutants year^{-1} and 19 036 t CO_2eq year^{-1}), yet modest when compared to overall city levels of air pollution and GHG emissions (2.66 % for PM_{10}, 0.43 % for NO_2, and 0.47 % for CO_2eq). Our research further shows that the effectiveness of green infrastructure-based strategies to meet environmental policy targets can vary greatly across pollutants. For example, our results suggest that NO_2 removal potential is unlikely to contribute in any substantial way to the compliance of current EU reference values. Therefore, for combating air pollution of NO_2, synergies between green infrastructure strategies and NO_2 emission curbing strategies (e.g., targeting road traffic) need to be searched and implemented in order to effectively deal with air quality regulations. On the other hand, PM_{10} removal potential should not be neglected in urban policy-making. Its contribution to the compliance with the current EU reference value can be substantial and potentially more effective than other local policies based on emissions abatement due to the importance of background pollution in Barcelona's PM_{10} levels.

Net carbon sequestration by urban forests has a very low influence when compared to total annual GHG city emissions, but our results suggest that it can contribute considerably to meet the 23 % GHG emissions reduction policy target until 2020, which only applies for emissions derived from services and activities directly managed by the City Council (2.10 % of total emissions).

We determine that the implementation of green infrastructure-based strategies at the municipal level (as is aimed by the *Barcelona Green*

Infrastructure and Biodiversity Plan 2020) would have a limited effect on local air quality levels and GHG emissions offsets, yet they would play a non-negligible complementary role to other policies intended to meet air quality (especially for PM_{10} levels) and climate change mitigation policy targets in Barcelona, fostering as well the provision of other important urban ecosystem services (e.g., urban temperature regulation, stormwater runoff mitigation, and recreational opportunities) at no additional monetary costs. We conclude that, in order to be effective, green infrastructure-based strategies to abate pollution in cities should be implemented at broader spatial scales (i.e., metropolitan area). However, it is critical that policy-makers consider an integrated approach in green infrastructure management, where possible trade-offs with other ecosystems services, disservices, and urban sustainability goals are fully acknowledged.

FOOTNOTES

1. Here "green space" corresponds to those areas with vegetation (e.g., urban parks, gardens, and other green areas) directly managed by the City Council. It includes also the natural and semi-natural areas of the Collserola Park, but it excludes green elements such as single street trees or private gardens.
2. "Green infrastructure is a concept addressing the connectivity of ecosystems, their protection and the provision of ecosystem services, while also addressing mitigation and adaptation to climate change" (EEA 2011).

REFERENCES

1. ASPB air quality report. 2011. Report on evaluation of the air quality in the city of Barcelona, year 2011. Agency for Public Health of Barcelona (ASPB), Barcelona, Spain, 75 pp (In Catalan).
2. Baik, J.J., K.H. Kwak, S.B. Park, and Y.H. Ryu. 2012. Effects of building roof greening on air quality in street canyons. Atmospheric Environment 61: 48–55.
3. Baldocchi, D.D., B.B. Hicks, and P. Camara. 1987. A canopy stomatal resistance model for gaseous deposition to vegetated surfaces. Atmospheric Environment 21: 91–101.
4. Barcelona City Council Statistical Yearbook. 2012. Department of Statistics, Barcelona City Council, Barcelona, Spain (In Spanish). Retrieved December 20, 2012, from http://www.bcn.cat/estadistica.

5. Barcelona Green Infrastructure and Biodiversity Plan 2020. 2013. Edited by Environmental and Urban Services. Urban Habitat Department. Barcelona City Council. Retrieved April 30, 2013, from http://www.bcn.cat/mediambient.
6. Beckett, K.P., P.H. Freer-Smith, and G. Taylor. 1998. Urban woodlands: Their role in reducing the effects of particulate pollution. Environmental Pollution 99: 347–360.
7. Bolund, P., and S. Hunhammar. 1999. Ecosystem services in urban areas. Ecological Economics 29: 293–301.
8. Burriel, J.A., J.J. Ibáñez, and J. Terradas. 2006. The ecological map of Barcelona, the changes in the city in the last three decades. XII National Spanish congress on geographic information technologies: University of Granada. ISBN 84-338-3944-6 (In Spanish, English summary).
9. Carter, W.P.L. 1994. Development of ozone reactivity scales for volatile organic compounds. Air & Waste 44: 881–899.
10. Chaparro, L., and J. Terradas. 2009. Report on Ecological services of urban forest in Barcelona. Barcelona City Council, Barcelona, Spain: Department of Environment. 96 pp.
11. Chiesura, A. 2004. The role of urban parks for the sustainable city. Landscape and Urban Planning 68: 129–138.
12. Currie, B., and B. Bass. 2008. Estimates of air pollution mitigation with green plants and green roofs using the UFORE model. Urban Ecosystems 11: 409–422.
13. Dobbs, C., F.J. Escobedo, and W.C. Zipperer. 2011. A framework for developing urban forest ecosystem services and goods indicators. Landscape and Urban Planning 99: 196–206.
14. Dodman, D. 2009. Blaming cities for climate change? An analysis of urban greenhouse gas emissions inventories. Environment and Urbanization 21: 185–201.
15. EC. 2008. Energy and climate package—Elements of the final compromise agreed by the European Council, European Commission. Retrieved December 15, 2012 from http://ec.europa.eu/clima/policies/package/documentation_en.htm.
16. EEA. 2011. Green infrastructure and territorial cohesion. The concept of green infraestructure and its integrations into policies using monitoring systems. EEA report 18/2011, 138. Copenghagen, Denmark: European Environment Agency.
17. EEA. 2013. Air quality in Europe—2013 report. EEA report 9/2013, European Environment Agency, Copenhagen, Denmark, 107 pp. ISBN 978-92-9213-406-8.
18. EPA. 2010. Technical Support Document: Social Cost of Carbon for Regulatory Impact Analysis Under Executive Order 12866. Interagency Working Group on Social Cost of Carbon, United States Government. Retrieved February 15, 2013, from http://www.epa.gov/climatechange/EPAactivities/economics/scc.html.
19. Escobedo, F.J., and D.J. Nowak. 2009. Spatial heterogeneity and air pollution removal by an urban forest. Landscape and Urban Planning 90: 102–110.
20. Escobedo, F.J., J.E. Wagner, D.J. Nowak, C.L. De la Maza, M. Rodríguez, and D.E. Crane. 2008. Analyzing the cost effectiveness of Santiago, Chile's policy of using urban forests to improve air quality. Journal of Environmental Management 86: 148–157.

21. Escobedo, F.J., S. Varela, M. Zhao, J.E. Wagner, and W. Zipperer. 2010. Analyzing the efficacy of subtropical urban forests in offsetting carbon emissions from cities. Environmental Science & Policy 13: 362–372.
22. Escobedo, F.J., T. Kroeger, and J.E. Wagner. 2011. Urban forests and pollution mitigation: Analyzing ecosystem services and disservices. Environmental Pollution 159: 2078–2087.
23. Fuller, R.A., and K.G. Gaston. 2009. The scaling of green space coverage in European cities. Biology Letters 5: 352–355.
24. Gómez-Baggethun, E., and D.N. Barton. 2013. Classifying and valuing ecosystem services for urban planning. Ecological Economics 86: 235–245.
25. Gómez-Baggethun, E., Å. Gren, D. Barton, J. Langemeyer, T. McPhearson, P. O'Farrell, E. Andersson, Z. Hamstead, et al. 2013. Urban ecosystem services. In Urbanization, ed. T. Elmqvist, M. Fragkias, J. Goodness, B. Güneralp, P. Marcotullio, R.I. McDonald, et al., 175–251. Challenges and Opportunities. Springer: Biodiversity and Ecosystem Services. ISBN 978-94-007-7087-4.
26. Hirabayashi, S., C.N. Kroll, and D.J. Nowak. 2011. Component-based development and sensitivity analyses of an air pollutant dry deposition model. Environmental Modelling and Software 26: 804–816.
27. Hoornweg, D., L. Sugar, and C.L. Trejos Gómez. 2011. Cities and greenhouse gas emissions: Moving forward. Environment and Urbanization 23: 207–227.
28. i-Tree User's manual. 2008. Tools for assessing and managing Community Forests. Software Suite v2.1. Retrieved November 15, 2012, from http://www.itreetools.org.
29. Jo, H.K., and G.E. McPherson. 1995. Carbon storage and flux in urban residential greenspace. Journal of Environmental Management 45: 109–133.
30. Kennedy, C., J. Steinberger, B. Gasson, Y. Hansen, T. Hillman, M. Havránek, D. Pataki, A. Phdungsilp, et al. 2009. Greenhouse gas emissions from global cities. Environmental Science & Technology 43: 7297–7302.
31. Kesselmeier, J., and M. Staudt. 1999. Biogenic volatile organic compounds (VOC): An overview on emission, physiology and ecology. Journal of Atmospheric Chemistry 33: 23–88.
32. Liu, C., and X. Li. 2012. Carbon storage and sequestration by urban forests in Shenyang, China. Urban Forestry & Urban Greening 11: 121–128.
33. Lyytimäki, J., and M. Sipilä. 2009. Hopping on one leg—The challenge of ecosystem disservices for urban green management. Urban Forestry & Urban Greening 8: 309–315.
34. Manning, W.J. 2008. Plants in urban ecosystems: Essential role of urban forests in urban metabolism and succession toward sustainability. International Journal of Sustainable Development and World Ecology 15: 362–370.
35. McHale, M.R., E.G. McPherson, and I.C. Burke. 2007. The potential of urban tree plantings to be cost effective in carbon credit markets. Urban Forestry & Urban Greening 6: 49–60.

36. McPherson, E.G., K.I. Scott, and J.R. Simpson. 1998. Estimating cost effectiveness of residential yard trees for improving air quality in Sacramento, California, using existing models. Atmospheric Environment 32: 75–84.
37. Murray, F.J., L. Marsh, and P.A. Bradford. 1994. New York state energy plan Vol. II: issue reports. Albany, NY: New York State Energy Research and Development Authority.
38. Nowak, D.J. 1994. Air pollution removal by Chicago's urban forest. In Chicago's Urban Forest Ecosystem: Results of the Chicago Urban Forest Climate Project, ed. E.G. McPherson, D.J. Nowak, and R.A. Rowntree, pp. 63–81. Radnor: USDA Forest Service General Technical Report NE-186.
39. Nowak, D.J. 2006. Institutionalizing urban forestry as a "biotechnology" to improve environmental quality. Urban Forestry & Urban Greening 5: 93–100.
40. Nowak, D.J., and D.E. Crane. 2000. The Urban Forest Effects (UFORE) Model: Quantifying urban forest structure and functions. In Integrated tools for natural resources inventories in the 21st century, ed. M. Hansen, and T. Burk, pp 714–720. St. Paul: North Central Research Station.
41. Nowak, D.J., and D.E. Crane. 2002. Carbon storage and sequestration by urban trees in the USA. Environmental Pollution 116: 381–389.
42. Nowak, D.J., K.L. Civerolo, S. Trivikrama Rao, G. Sistla, C.J. Luley, and D.E. Crane. 2000. A modeling study of the impact of urban trees on ozone. Atmospheric Environment 34: 1601–1613.
43. Nowak, D.J., D.E. Crane, J.C. Stevens, and M. Ibarra. 2002a. Brooklyn's Urban Forest, 107 pp. Newtown Square, PA: USDA Forest Service, Northeastern Research Station, GTR NE-290.
44. Nowak, D.J., J.C. Stevens, S.M. Sisinni, and C.J. Luley. 2002b. Effects of urban tree management and species selection on atmospheric carbon dioxide. Journal of Arboriculture 28: 113–122.
45. Nowak, D.J., D.E. Crane, and J.C. Stevens. 2006. Air pollution removal by urban trees and shrubs in the United States. Urban Forestry & Urban Greening 4: 115–123.
46. Nowak, D.J., D.E. Crane, J.C. Stevens, R.E. Hoehn, and J.T. Walton. 2008a. A ground-based method of assessing urban forest structure and ecosystem services. Arboriculture & Urban Forestry 34: 347–358.
47. Nowak, D.J., J.T. Walton, J.C. Stevens, D.E. Crane, and R.E. Hoehn. 2008b. Effect of plot and sample size on timing and precision of urban forest assessments. Arboriculture & Urban Forestry 34: 386–390.
48. Oberndorfer, E., J. Lundholm, B. Bass, R.R. Coffman, H. Doshi, N. Dunnett, S. Gaffin, M. Köhler, et al. 2007. Green roofs as urban ecosystems: Ecological structures, functions, and services. BioScience 57: 823.
49. Paoletti, E. 2009. Ozone and urban forests in Italy. Environmental Pollution 157: 1506–1512.
50. Pataki, D.E., P.C. Emmi, C.B. Forster, J.I. Mills, E.R. Pardyjak, T.R. Peterson, J.D. Thompson, and E. Dudley-Murphy. 2009. An integrated approach to improving fossil fuel emissions scenarios with urban ecosystem studies. Ecological Complexity 6: 1–14.

51. Pataki, D.E., M.M. Carreiro, J. Cherrier, N.E. Grulke, V. Jennings, S. Pincetl, R.V. Pouyat, T.H. Whitlow, et al. 2011. Coupling biogeochemical cycles in urban environments: Ecosystem services, green solutions, and misconceptions. Frontiers in Ecology and the Environment 9: 27–36.
52. Pauleit, S., N. Jones, G. Garcia-Martin, J.L. Garcia-Valdecantos, L.M. Rivière, L. Vidal-Beaudet, M. Bodson, and T.B. Randrup. 2002. Tree establishment practice in towns and cities—Results from a European survey. Urban Forestry & Urban Greening 1: 83–96.
53. PECQ. 2011. The energy, climate change and air quality plan of Barcelona (PECQ) 2011–2020. Barcelona City Council. Retrieved December 15, 2012, from http://www.covenantofmayors.eu/about/signatories_en.html?city_id=381&seap.
54. Peñuelas, J., and J. Llusià. 2003. BVOCs: Plant defense against climate warming? Trends in Plant Science 8: 105–109.
55. Pérez, L., J. Sunyer, and N. Künzli. 2009. Estimating the health and economic benefits associated with reducing air pollution in the Barcelona metropolitan area (Spain). Gaceta sanitaria/SESPAS 23: 287–94 (In Spanish, summary in English).
56. Pons, X. 2006. MiraMon. Geographic information system and remote sensing software. Centre for Ecological Research and Forestry Applications (CREAF). ISMB: 84-931223-5-7.
57. Pugh, T.A.M., A.R. Mackenzie, J.D. Whyatt, and C.N. Hewitt. 2012. Effectiveness of green infrastructure for improvement of air quality in urban street canyons. Environmental Science and Technology 46: 7692–7699.
58. Robinson, M.F., J. Heath, and T.A. Mansfield. 1998. Disturbances in stomatal behaviour caused by air pollutants. Journal of Experimental Botany 49: 461–469.
59. Rowe, D.B. 2011. Green roofs as a means of pollution abatement. Environmental Pollution 159: 2100–2110.
60. Seibert, P., F. Beyrich, S.E. Gryning, S. Joffre, A. Rasmussen, and P. Tercier. 2000. Review and intercomparison of operational methods for the determination of the mixing height. Atmospheric Environment 34: 1001–1027.
61. Setälä, H., V. Viippola, A.L. Rantalainen, A. Pennanen, and V. Yli-Pelkonen. 2013. Does urban vegetation mitigate air pollution in northern conditions? Environmental Pollution 183: 104–112.
62. Taha, H. 1996. Modeling impacts of increased urban vegetation on ozone air quality in the South Coast Air Basin. Atmospheric Environment 30: 3423–3430.
63. Terradas, J., T. Franquesa, M. Parés, and L. Chaparro. 2011. Ecología urbana. Investigación y Ciencia 422: 52–58. (In Spanish).
64. U.S. Department of Labor. Bureau of Labor Statistics. Retrieved January 15, 2013, from http://www.bls.gov/ppi/.
65. Vos, P.E.J., B. Maiheu, J. Vankerkom, and S. Janssen. 2013. Improving local air quality in cities: To tree or not to tree? Environmental Pollution 183: 113–122.
66. Yang, J., J. McBride, J. Zhou, and Z. Sun. 2005. The urban forest in Beijing and its role in air pollution reduction. Urban Forestry & Urban Greening 3: 65–78.
67. Yang, J., Q. Yu, and P. Gong. 2008. Quantifying air pollution removal by green roofs in Chicago. Atmospheric Environment 42: 7266–7273.

68. Zhao, M., Z. Kong, F.J. Escobedo, and J. Gao. 2010. Impacts of urban forests on offsetting carbon emissions from industrial energy use in Hangzhou, China. Journal of Environmental Management 91: 807–813.

CHAPTER 2

Modelling Short-Rotation Coppice and Tree Planting for Urban Carbon Management–A Citywide Analysis

NICOLA MCHUGH, JILL L. EDMONDSON, KEVIN J. GASTON, JONATHAN R. LEAKE, AND ODHRAN S. O'SULLIVAN

2.1 INTRODUCTION

Urban populations depend on rural areas to supply essential provisioning ecosystem services including food, fibres, wood and water, and it is often assumed that urban areas are unable to make any significant contribution to such services. However, urban greenspaces deliver a variety of supporting, regulating and cultural ecosystem services (Davies et al. 2011a; Gómez-Baggethun et al. 2013; Nowak et al. 2013a), including high species richness (McKinney 2008), improved psychological well-being (Fuller et al. 2007), reduced stormwater run-off and air pollution interception (Sæbø et al. 2012). Better management of urban greenspace to deliver multiple ecosystem services has the potential to simultaneously enhance the quality of life for city dwellers and the sustainability of urban areas (Davies et al. 2011a). Despite such evidence, the potential for urban greenspaces to deliver provisioning ecosystem services such as biomass fuel and timber, and regulating services, such as carbon storage, has received little attention in the UK. Consequently, the extent to which tree planting can contribute to CO_2 emissions reduction targets through carbon sequestration into biomass or through biofuel substitution for fossil fuels in UK cities remains unclear.

Urban areas are expanding globally, with urban populations increasing fivefold from 0·8 to 3·6 billion between 1950 and 2011 (United Nations 2012), and these areas disproportionately contribute to global anthropogenic CO_2 emissions (UN-Habitat 2011). The UK is committed to reducing national CO_2 emissions by 80% of 1990 values by 2050 (UK Parliament 2008), requiring a major reduction in fossil fuel use. Maximizing local energy production and increasing carbon sequestration into biomass will undoubtedly be among the range of solutions required to achieve this ambitious goal.

Appropriately planned and managed, urban greenspaces could deliver increases in specific ecosystem services such as carbon storage in trees, as seen in urban tree planting in the UK (Díaz-Porras, Gaston & Evans 2014) and USA (Nowak et al. 2013b; McPherson & Kendall 2014). In Leicester, a typical UK city, trees account for 97·3% of carbon stored in above-ground vegetation (Davies et al. 2011b) confirming their importance in ecosystem carbon storage. Urban tree planting has been promoted to enhance multiple ecosystems service benefits (Roy, Byrne & Pickering 2012) including: air pollution interception (Sæbø et al. 2012); noise reduction (Roy, Byrne & Pickering 2012); enhanced stormwater infiltration (Stovin, Jorgensen & Clayden 2008); reduced building energy use for summer cooling (Rahman, Armson & Ennos 2014) and recreation, aesthetic and cultural benefits (Kaplan 2007).

Larger greenspace areas may have the potential for growing short-rotation coppice (SRC), a system for woody biomass production. SRC refers to any woody species (typically high-yielding species such as poplar and willow), which is managed in a coppice system, typically harvested every 3–5 years and normally grown as a biofuel crop (Aylott et al. 2008, 2010). This can contribute to the UK Government target for 15% of energy to come from renewable sources by 2020 (DECC 2011).

Despite the large areas of greenspace within towns and cities, current UK SRC guidance is exclusively focussed on agricultural land (Natural England, 2013a). However, constraints identified in this guidance do not necessarily preclude SRC in urban areas, indeed the urban fringe was identified as particularly suited to such crops in an earlier report (British BioGen 1996). Many of the recommendations for increasing biodiversity within SRC patches (Rowe, Street & Taylor 2009) are achievable in urban areas, including plantations with large edge to interior ratio, small plot sizes and blocks of SRC interspersed with other habitats.

The fragmented heterogeneous structure of urban landscapes due to division of land into small patches under different ownership, management and diverse usage (Luck & Wu 2002) is exemplified by domestic gardens which account for 22–27% of greenspace in UK urban areas (Loram et al. 2007). High-resolution spatial data are overcoming the problem of assessing the ecosystem services provided by such small land parcels (Davies et al. 2013).

Here, we assess the potential to increase carbon sequestration in trees and harvested SRC biomass in a typical UK city. On the basis of previous estimates, the contribution of SRC biomass to heat municipal buildings and homes and the reduction in CO_2 emissions achieved by this biomass substituting for natural gas heating homes is assessed. Wood-fuel biomass boilers have gained increasing importance in municipal heating systems and schools (The Carbon Trust, 2012); however, there has been surprisingly little research to date on biomass fuel production in urban areas (but see Nielsen & Møller 2008; MacFarlane 2009; Strohbach et al. 2012; McPherson & Kendall 2014; Zhao et al. 2014).

We developed modelling tools to address the specific challenges of simulating tree and SRC growth to ensure that the modelled trees could be fitted into the existing landscape and continue to do so as they grew. The tree-planting model identified suitable sites for planting and was designed to maintain the existing diversity of tree species within the urban study area, based on recent surveys of trees in Leicester (Davies et al. 2011b), matching tree size at maturity to the greenspace patch sizes.

2.2 MATERIALS AND METHODS

2.2.1 STUDY AREA

This study focused on Leicester (52°38′N, 1°08′W), a typical mid-sized city in central England with a population of around 310 000, and annual CO2 emissions of 478 000 tonnes of carbon (Leicester City Council, 2012). The 73-km^2 city area has a densely developed urban core, beyond which are suburbs, with built development reaching the city boundary in the east and west and small peri-urban areas to the north and south. The annual daily mean temperature range is 1·7–21·3 °C with 606-mm annual rainfall (Met Office 2012).

Land ownership was divided into private (land within the boundary of private dwellings, identified through MasterMap) (Ordnance Survey 2008), public (land owned by Leicester City Council) or mixed-land ownership (areas belonging to business or private individuals and land where ownership was undetermined). Land cover was derived from the Land-Base data set (Infoterra 2006), which identifies eight land cover classes: bare ground, inland water, artificial surface, buildings, herbaceous (mainly grassland), shrub, tall shrub and trees (0·25 m^2 resolution). Only areas categorized as herbaceous or bare ground were considered suitable for tree or SRC planting in our models, with shrub, tall shrub and tree land cover, and areas currently under artificial surface or buildings, excluded.

2.2.2 MIXED-SPECIES TREE-PLANTING MODELS

Separate mixed-species tree-planting models were developed to apply to private land (Fig. S1, Supporting information) and public and mixed ownership land (Fig. S2), as the small land parcel size in private land necessitated the use of a separate model. The two GIS models (ESRI ArcInfo 10, ModelBuilder) iteratively planted trees allowing planting restriction to be applied to avoid areas deemed unsuitable (Table S2).

Building on an approach developed by Wu, Xiao & McPherson (2008) for Los Angeles, the models analysed the current landscape in order to predict the ability to accommodate trees, including allowing for tree growth over 25 years, a modelling time span that reflects the use of current climate information and is consistent with recent studies of effects of peri-urban trees on air quality (Kroeger et al. 2014). Combining data from the tree survey carried out by Davies et al. (2011b) and a garden tree survey using the same methodology (data available from the Dryad Digital Repository: http://dx.doi.org/10.5061/dryad.j25t0; McHugh et al. 2015), over 1300 trees in Leicester were identified and diameter at breast height (d.b.h.) measured. Those species with more than one individual (68 species) were included in the tree-planting models.

Mature crown diameter values of large (15 m) and small (5 m) species within the tree population were incorporated into the models reducing the risk of overplanting the landscape, replicating the species and size heterogeneity of the current urban forest and developing more realistic carbon storage values than could be achieved with a single species planting model. Trees planted were modelled on whips [<2 cm diameter, 100–200 cm height (ENA 2010; Forestry Commission 2010)], with a mean diameter planting size of 0·53 cm determined from Willoughby et al. (2007).

Minimum distance restrictions from impervious surfaces (measured from trunk) of 6 or 2 m for large and small trees, respectively, were applied. These values were determined by combining root spread values of tree species from the local population, expressed as a percentage of mean crown diameter (Gruffydd 1987; Hodge & White 1990; RHS 2014), together with existing distance guidelines to minimize damage to nearby buildings, roads and paths (Gasson & Cutler 1998) (Table S1). Such guidelines have economic relevance—in the London Borough of Hackney, UK, 40% of trees removed from 2002 to 2007 were a result of insurance claims for tree-related property damage (LAEC, 2007).

The private ownership model (Fig. S1) in domestic gardens had a minimum area requirement of 9 m^2 for large trees and 2 m^2 for small trees with no overlap of existing or newly planted tree canopies stipulated. The model continued searching for planting sites until the number of trees planted in each cycle was <10 large or 1000 small trees, determined to balance search time with additional trees planted. The separate modelling approach applied to public and mixed ownership land was designed to maximize planting in larger spaces (Fig. S2). This model incorporated a single cycle of large tree planting followed by the removal of unsuitably sited trees, that is where mature canopies would extend beyond the suitable planting area. The final stage identified sites that could still accommodate small trees and filled gaps within the planting scheme. Identical tree size and minimum distances to buildings, roads and paths were used in private, and public and mixed ownership models.

Urban-specific mortality rates for newly planted trees (0–3 years) of 10%, and for established trees (4–25 years) of 6%, were applied (Gilbertson & Bradshaw 1990; Nowak, McBride & Beatty 1990; Bradshaw, Hunt & Walmsley 1995; Nowak, Kuroda & Crane 2004; LAEC 2007). A replanting phase (5% trees aged 0–3 years, 3% trees aged 4–25 years) then occurred outside the spatial modelling environment. The number and size of trees removed from the models through annual mortality events was calculated in order to quantify carbon removed from the study area.

Annual tree growth rates were taken from the literature and applied for 25 years to planted trees. Species-specific rates were used when available, or else genus or family specific rates were used (see Table S3), with growth rates of urban trees in the same geographic region as the study site used preferentially. Linear growth rates were applied as growth is unlikely to slow in the first 25 years (Strohbach et al. 2012). The aboveground biomass of trees was calculated annually using species- and genus-specific allometric equations (see Table S4), and a biomass-to-carbon conversion factor of 0·46 for broadleaf and 0·42 for coniferous species was used to determine carbon content (Milne & Brown 1997). The use of generalized equations (up to eight annual growth rates and six allometric biomass equations) minimized variability, an issue identified by McHale et al. (2009) when applying non-urban equations to urban trees. To compare the mixed-species models, the maximum possible increase in carbon storage by tree planting was estimated using the fastest growing large (*Eucalyptus gunnii* Hook. F.) and small trees (*Populus tremula* L.) in our data base (Table S3).

2.2.3 SRC MODEL

Potential SRC yield for combined willow and poplar plantings was calculated based on regional mean values based on Agricultural Land Classification (ALC) (Aylott et al. 2010). As no yield value was provided for the ALC 'urban' category, the yield for lowest quality (category 5) land, of 10·3 oven-dry tonnes (odt) ha^{-1} $year^{-1}$, was used. This is a conservative approach as citywide analysis of soil properties in Leicester found that in most greenspaces, the soil quality matches or exceeds that of agricultural land (Edmondson et al. 2011, 2012, 2014). A series of spatial restriction criteria, based on UK Energy Crop Scheme guidance (Natural England 2013b) and findings of biofuels research (Renewable Fuels Agency 2008; Aylott et al. 2010), was developed (Table S2) to identify suitable planting sites and the annual yield possible across the study area was calculated. The heating and fossil fuel offset potential of SRC yields were estimated (see Appendices S1 and S2) using published values for the biomass of wood chips required to heat a typical domestic house, municipal building or support a district heating scheme (Biomass Energy Centre 2014). The fossil fuel carbon savings of biomass substitution for natural gas was calculated using data on household gas consumption from DECC (2013), and the net fossil fuel savings relative to natural gas provided by SRC wood chips, taking into account fossil fuel costs of harvesting, transport, chipping, drying and distribution (Defra 2009).

2.2.4 COMPARISON OF TREE AND SRC PLANTING MODEL OUTPUTS

The increase in carbon sequestration resulting from the two carbon management approaches, the mixed-species tree planting and SRC models was compared at years 10 and 25 to the above-ground carbon stocks of the existing tree population of the study area. In addition, a combined management approach giving priority to SRC on all suitable land followed by the application of the mixed-species tree-planting model to remaining suitable sites was employed to maximize effects of carbon management.

2.3 RESULTS

The tree-planting models identified an area of 11 km^2 suitable for planting, 86·5% of which was in public or mixed ownership, and only 13·5% was in private gardens (Table 1). Nonetheless, gardens were found to be able to accommodate 70 000 additional, mainly small, trees. Over 25 years, these trees could enhance carbon stocks by six times the current amounts in above-ground herbaceous vegetation in the areas of gardens allocated to tree planting (Tables 1 and 2). This is a higher proportional increase in carbon storage than that found by the model of public or mixed ownership land, which projects a doubling of carbon storage over 25 years in areas of herbaceous vegetation allocated to the planting of a total of 220 000 trees. Most of these trees were of species too large for gardens once fully grown and therefore were planted at a lower density than the small trees.

Carbon storage increases resulting from applying the tree-planting models are strongly influenced by the differing tree species compositions between land ownership classes. On domestic land, 23% of trees were fast-growing *Cupressaceae* which over the 25-year period individually sequestered c. 96-kg carbon (d.b.h. 33 cm). The species composition of trees found in public and mixed ownership land was more diverse and although the most common tree species have the potential to reach a large size, they often grow more slowly, for example *Fraxinus excelsior* L. with a d.b.h. of 14 cm at 25 years. Because of the initially small size and associated slow growth rates of many of the trees, the model projected a total increase in above-ground carbon storage in biomass compared to herbaceous vegetation by only 2600–4200 tonnes over 25 years (Tables 1 and 2). However, as a consequence, we expect tree planting to supplement rather than to replace the existing herbaceous biomass. Carbon removed from the study area as a result of tree mortality over 25 years totalled 224 tonnes of carbon (private land ownership model) and 460 tonnes of carbon (public and mixed ownership model), giving a total removal of tree biomass of 684 tonnes. Although likely to be unacceptable from a biodiversity and aesthetic perspective (Roy, Byrne & Pickering 2012), maximizing carbon sequestration using the fastest growing large and small tree species (*E. gunnii* and *P. tremula*) indicated potential increased storage of 53 000

Table 1. Area of greenspace suitable for tree planting or short-rotation coppice (SRC), and estimates of the above-ground carbon stocks in vegetation in these areas

Greenspace management approach	Land ownership	Total greenspace area under herbaceous vegetation (m^2)	Area of herbaceous greenspace suitable for management approach[a]		Current above-ground carbon in area suitable for management approach[b] (tonnes)
			m^2	%	
Tree planting	Public	12 647 614	3 096 813	47·5	464·522
	Mixed	6 524 299	6 475 435	51·2	906·561
	Private	8 402 581	1 494 506	17·8	209·231
	All	27 574 494	11 066 754	40·1	1580·314
SRC establishment	Public	12 647 614	1 710 878	26·2	256·632
	Mixed	6 524 299	4 154 263	32·8	581·597
	All	19 171 913	5 865 141	30·6	838·229
Combined	All	27 574 494	11 066 754	40·1	1580·314

[a] Suitable areas were identified after spatial restriction criteria were applied (areas covered in shrubs or trees were excluded).
[b] See Davies et al. (2011b) for further details.

tonnes of carbon after 25 years—over 12 times greater than the projection from the model with multiple species (Table 2).

In comparison with tree planting, the SRC planting model projected much larger total biomass production of 71 848 tonnes across the city over 25 years, 20 958 tonnes of carbon being produced by SRC on public land and 50 889 tonnes of carbon on mixed ownership land (Table 2). These quantities are striking considering that the SRC model identified only 5·87 km2 (8% of the city) as suitable for planting, reflecting the high planting density and repeated harvesting of fast-growing coppice biomass every 4 years which allows for rapid regrowth and associated conversion of atmospheric carbon to biomass.

Under the combined tree planting and SRC management, 73 400 tonnes of extra carbon could be captured by tree biomass and harvested SRC biomass (Tables 1 and 2) using 15% of the land area across Leicester. Total carbon removed by tree mortality in this case was estimated to be only 245 tonnes of carbon over 25 years.

The spatial distribution of current above-ground carbon in Leicester, together with projected 25-year carbon conversion to live biomass (trees) and harvested biomass (SRC), is presented in Fig. 1. Current stocks of above-ground carbon (Fig. 1a) average 3·16 kg m^{-2}, with greatest storage corresponding with managed parkland and other large greenspaces, largely on the city outskirts. Under the tree-planting approach (Fig. 1b), increases are rarely above 0·06 kg of carbon m^{-2} in the city centre after 25 years owing to lack of space for large trees. Outside the city centre, a higher proportion of land is suitable for tree planting, but our models show across the city above-ground carbon stocks only increase by 0·04–3·20 kg m^{-2} after 25 years. Nonetheless, these increases should be viewed in the context of the already high biomass of vegetation in the city compared to the UK average above-ground vegetation carbon density of 0·497 kg carbon m^{-2} (Milne & Brown 1997).

The areas suitable for SRC establishment are more limited and mainly in the urban fringes (Figs 1c and 2a). However, it is clear that where land is suitable for SRC, the quantity of carbon that can be fixed is far greater than that achievable by planting trees using a mixture of species similar to the existing urban tree population (Fig. 1b,c; Table 2).

Table 2. Potential increase in carbon sequestration into live trees and harvested short-rotation coppice (SRC) biomass over 25 years, and potential carbon offsetting by SRC biomass substitution for natural gas in domestic heating and tree planting

Greenspace management approach	**Carbon (tonnes) sequestered into newly planted trees or harvested SRC biomass [carbon offset by SRC, and under combined management the total carbon sequestered plus offset for tree planting plus SRC]**		
	Year 0[a]	Year 10	Year 25
Tree planting			
Public ownership	0·286	167·377	1024·389
Mixed ownership	0·512	294·266	1821·020
Private ownership	7·226	249·024	1337·278
Total	8·024	710·667	4182·687
SRC establishment			
Public ownership	0	8383·302 [3411·341]	20958·256 [8528·354]
Mixed ownership	0	20355·889 [8283·238]	50889·722 [20708·096]
Total	0	28739·191 [11694·580]	71847·978 [29236·450]
Combined management approach	7·726	29309·877 [12405·247]	74983·920 [33419·137]

[a] Year 0 values refer to imported carbon for tree-planting establishment. The carbon import of SRC is assumed to be zero as establishment is from small cuttings.

The spatial distribution of potential carbon capture into trees and harvested SRC biomass production (Fig. 2b) clearly identifies areas, primarily on the city margins, with the greatest opportunities for a change in management. These are larger patches of public parks, undeveloped greenspace and brownfield sites near to industrial zones. The largest increases are due primarily to SRC, but enhancement of carbon stocks can take place across most of the city through utilizing small patches of urban greenspace for tree planting.

Based on our modelled SRC biofuel production potential across the city, averaging these yields over 25 years, could supply energy to 30 municipal buildings, or 52 district heating schemes (common in northern Europe and well suited to densely populated urban areas) (Biomass Energy Centre 2014). Using data from an award-winning scheme in Barnsley, UK (Barnsley Metropolitan Borough Council 2006), the SRC biomass could support district heating of over 4200 flats, comprising 3% of households in Leicester. Domestic use of woodchip biofuel from SRC for heating would allow 1566 households to each avoid emissions of 746·7 kg carbon $year^{-1}$ compared to the use of fossil fuel natural gas (Defra 2009), potentially avoiding 29 236 tonnes of fossil fuel carbon release over 25 years (Table 2). Together with the carbon sequestration into trees, additional to pre-existing herbaceous vegetation, a total reduction of 33 419 tonnes of carbon in the atmosphere could be achieved in 25 years by combined SRC and tree planting across the city (Table 2).

2.4 DISCUSSION

The analysis presented here highlights the potential for enhanced carbon storage and mitigation of anthropogenic CO_2 emissions by tree planting and SRC in urban greenspaces in a typical UK city. Assessment of carbon accumulation in urban tree-planting programmes is constrained by the limited availability of urban-specific tree growth data. Our models mostly used growth rates reported for Europe (67%) (Table S4) and North America (13%) (Table S4). Urban-specific growth rates only accounted for 4% of those used, reflecting the limited availability of these data. Most growth rates were derived from community woodland (24%),

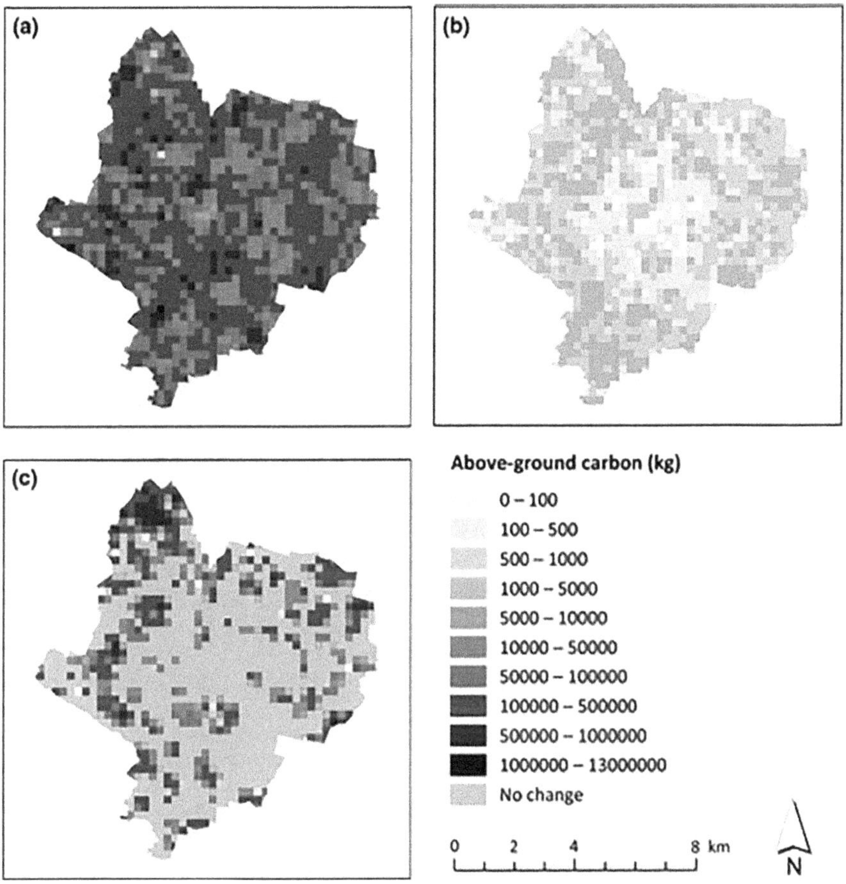

FIGURE 1: (a) Current total above-ground carbon in 250 Ч 250 m grids across the city, (b) additional biomass carbon after 25 years predicted by the mixed-species tree-planting models and (c) carbon converted to harvested biomass over 25 years predicted by the short-rotation coppice (SRC) model.

forestry (22%) and ex-agricultural (16%) sites. The application of natural forest system allometric relationships to urban forests is commonplace (Timilsina et al. 2014), but potentially inaccurate. However, our use of averaged equations is one method of constraining errors in biomass estimates (McHale et al. 2009).

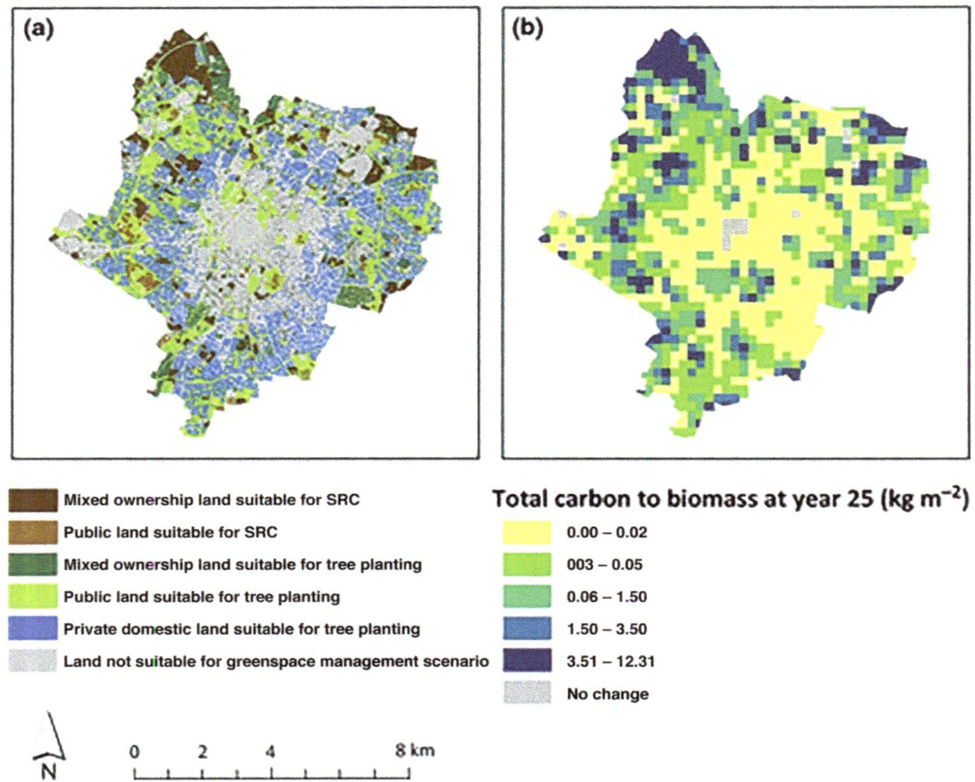

FIGURE 2: (a) Available urban greenspace suitable for management under the combined management approach and (b) total carbon assimilated both into above-ground tree biomass, and harvested in short-rotation coppice (SRC) over 25 years under the combined management approach in 250 Ч 250 m grids.

Fossil fuel carbon emissions occur in the nursery-raising, transport, and planting of new trees and their subsequent maintenance (Nowak & Crane 2002; Strohbach et al. 2012; McPherson & Kendall 2014). These emissions are very context dependant. In the Million Trees Los Angeles Programme which covers an area of 1022 km^2, McPherson & Kendall (2014) estimate that 6·8 kg of fossil fuel carbon is required to grow and plant each tree, mainly through use of oil in transport. In the more compact UK cities, these carbon costs are likely to be much lower. The modelled fitting of trees to suitable-sized patches in our study results in low planting densities that will minimize the need for maintenance over 25 years. Furthermore, a comparable study of urban tree planting found the majority of trees did not need pruning (Russo et al. 2014), and McPherson & Kendall (2014) suggest urban tree maintenance is only about 3% of the net reduction in CO_2 due to tree planting arising from sequestration into biomass and avoided fossil fuel carbon emissions where harvest biomass is used as a biofuel.

If our findings in Leicester are representative of the 6·8% of the UK that is urban area (Davies et al. 2011a), 15% of this land is suitable for combined planting of SRC and trees, suggesting that these areas hold the potential for reducing fossil fuel carbon emissions and increasing tree carbon sequestration by a total of over 7 480 000 tonnes carbon over 25 years nationally. This is a first approximation, assuming SRC is used to substitute natural gas in domestic heating, and is based on 10·3 odt ha^{-1} $year^{-1}$ SRC yield (Aylott et al. 2010), rather than the 6 odt ha^{-1} year−1 value of Strohbach et al. (2012). In Leicester, soil quality data (Edmondson et al. 2011, 2012, 2014) justify the higher yield value. More definitive estimates of carbon savings require the tree and SRC yields on typical urban soils and landscapes to be determined, and the areas of urban land suitable for planting to be determined nationally.

Short-rotation coppice biofuel production requires fossil fuel energy use by machinery for planting, management, harvesting and processing, resulting in carbon emissions estimated to be c. 22% of the total global warming potential of SRC biofuel in the Mediterranean (Esteban et al. 2014). These components have been estimated for UK SRC production by Defra (2009) and are taken into account in our calculations of avoided carbon emissions, but are not based on urban grown SRC. In an urban context, data are required on land-use change effects on other greenhouse gasses such as N_2O (Don et al. 2012) and a life cycle assessment made of the transport and processing activities (St Clair, Hillier & Smith 2008; Holtsmark 2013). Local production and consumption will minimize transport emissions, estimated to be 11·5% of the global warming potential of SRC biofuel production in a Spanish case study (Esteban et al. 2014), increasing the economic viability for district energy schemes (Climate East Midlands 2012).

To meet the UK government target of 15% of all energy and 30% of electricity demand to come from renewable sources by 2020 (DECC 2009), Aylott et al. (2010) calculate 0·8 million ha would be required if met by SRC production. To achieve the 7·5 million odt required, all grade 5 and 97% of grade 4 agricultural land across England would be needed to avoid the best quality land. SRC production across England from 2010 to 2011 ranged from 2600 to 2700 ha (Defra 2013), indicating low acceptance of SRC by farmers. Our modelling suggests it is possible to add over 20% to the current UK SRC output by utilizing urban sites within Leicester alone. Assuming Leicester is not unique, our findings underline the untapped potential for SRC across UK urban areas.

The greatest potential for an enhanced urban carbon sequestration strategy is on the urban fringe, comprising predominantly public and mixed ownership land that can be used for tree planting or SRC. However, changed greenspace management over large areas of the city has implications for existing and future provision of ecosystem services. Urban tree planting is recognized to improve local provision of ecosystem services in ways that can positively influence local climate, carbon cycles and energy use (Davies et al. 2011b; Nowak et al. 2013a). The establishment of SRC would allow for increases in pollutant interception, microclimate amelioration, soil stabilization, visual amenity additions to heterogeneous urban areas and provide graded edges to forested areas (Wiström et al. 2015). However, SRC could negatively impact local ecosystem services potentially restricting public access to greenspaces and may have low public acceptance in some areas owing to the episodic aesthetic contrasts between dense mature coppice and recently harvested stools (Nielsen & Møller 2008). It is important that factors such as these are taken into consideration when selecting suitable sites for any energy crop (Aylott et al. 2010; Bullock et al. 2011). Plantations on transport route embankments may have noise reduction and pollution interception benefits, although the need for buffer zones and access for harvesting and management may ultimately exclude such sites. This highlights the importance of identifying competing interests of stakeholders, as conflicts may arise if single ecosystem services are promoted in isolation to the wider consequences (Bullock et al. 2011). Large areas of many cities are former industrial and derelict building, brownfield sites that are often contaminated, requiring expensive remediation before redevelopment. Such sites naturally support invading pioneer trees and could support SRC, with the added benefit of soil phytoremediation (French et al. 2006) although, when burning biomass, appropriate filters would need to be used (Zhao et al. 2014).

In conclusion, this study highlights the potential of urban greenspace for enhanced carbon management through SRC and tree planting. Carbon sequestration benefits from tree planting would continue well beyond the 25-year scope of this study, as older trees disproportionately contribute to carbon storage (Davies et al. 2011b). In contrast, the benefits from fossil fuel replacement by SRC are realized much sooner, with just one mid-sized city having the potential to add over 20% to UK production of this biomass fuel in about a decade. Even if cities across the UK only implemented a portion of the combined management approach suggested in this study, the potential for increased SRC production could reduce demand for high-quality agricultural land to be used for biofuel production and its associated loss of food production (Renewable Fuels Agency 2008), with potential economic and societal benefits. Local authorities are central to national efforts to cut greenhouse gas emissions and need to encourage the use of urban spaces to assist in meeting the 80% reduction in CO_2 emissions by 2050 target (UK Parliament 2008) and the EU target of 20% renewable energy by 2020 (DTI, DFT & DEFRA, 2007). The development of biomass energy sources close to large populations and encouragement of landowners (public and private) to increase carbon sequestration across a city should be part of climate change mitigation policies of city councils.

REFERENCES

1. Aylott, M.J., Casella, E., Tubby, I., Street, N.R., Smith, P. & Taylor, G. (2008) Yield and spatial supply of bioenergy poplar and willow short-rotation coppice in the UK. New Phytologist, 178, 358–370.
2. Aylott, M.J., Casella, E., Farrall, K. & Taylor, G. (2010) Estimating the supply of biomass from short-rotation coppice in England, given social, economic and environmental constraints to land availability. Biofuels, 1, 719–727.
3. Barnsley Metropolitan Borough Council (2006) District heating from local tree waste. Technical report. Available at: www.ashden.org/winners/barnsley (accessed April 2015).
4. British BioGen (1996) Short Rotation Coppice for Energy Production. British BioGen & Department for Trade and Industry, London.
5. Biomass Energy Centre (2014) Biomass heating of buildings of different sizes. Available at: www.biomassenergycentre.org.uk/portal/page?_pageid=75,163211&_dad=portal&_schema=PORTAL (accessed April 2015).
6. Bradshaw, A.D., Hunt, B. & Walmsley, T. (1995) Trees in the Urban Landscape. E & F N Spon, London.
7. Bullock, J.M., Aronson, J., Newton, A.C., Pywell, R.F. & Rey-Benayas, J.M. (2011) Restoration of ecosystem services and biodiversity: conflicts and opportunities. Trends in Ecology & Evolution, 26, 541–549.
8. Climate East Midlands (2012) District Heating in Leicester. Leicester City Council, Leicester. Available at: http://www.climate-em.org.uk/images/uploads/CEM-Leicester-DH-7-A4.pdf.
9. Davies, L., Kwiatkowski, L., Gaston, K.J., Beck, H., Brett, H., Batty, M. et al. (2011a) Urban. UK National Ecosystem Assessment: Technical Report, 361–410.
10. Davies, Z.G., Edmondson, J.L., Heinemeyer, A., Leake, J.R. & Gaston, K.J. (2011b) Mapping an urban ecosystem service: quantifying above-ground carbon storage at a city-wide scale. Journal of Applied Ecology, 48, 1125–1134.
11. Davies, Z.G., Dallimer, M., Edmondson, J.L., Leake, J.R. & Gaston, K.J. (2013) Identifying potential sources of variability between vegetation carbon storage estimates for urban areas. Environmental Pollution, 183, 133–142.
12. DECC (2009) The UK Renewable Energy Strategy 2009. Department of Energy & Climate Change, London.
13. DECC (2011) UK Renewable Energy Roadmap. Department of Energy & Climate Change, London.
14. DECC (2013) Energy Consumption in the UK (2013). Department of Energy & Climate Change, London.
15. Defra (2009) Carbon Factor for Wood Fuels for the Supplier Obligation. AEA Final Report to Department for Environment Food and Rural Affairs, London.
16. Defra (2013) Area of crops grown for bioenergy in England and the UK: 2008–2011. Available at: http://www.defra.gov.uk/statistics/foodfarm/landuselivestock/nonfoodcrops/ (accessed April 2015).
17. Díaz-Porras, D.F., Gaston, K.J. & Evans, K.L. (2014) 110 Years of change in urban tree stocks and associated carbon storage. Ecology and Evolution, 4, 1413–1422.

18. Don, A., Osborne, B., Hastings, A., Skiba, U., Carter, M.S., Drewer, J. et al. (2012) Land-use change to bioenergy production in Europe: implications for the greenhouse gas balance and soil carbon. GCB Bioenergy, 4, 372–391.
19. DTI, DFT & DEFRA (2007) UK Biomass Strategy. Department for Environment, Food and Rural Affairs, London.
20. Edmondson, J.L., Davies, Z.G., McCormack, S.A., Gaston, K.J. & Leake, J.R. (2011) Are soils in urban ecosystems compacted? A citywide analysis. Biology Letters, 7, 771–774.
21. Edmondson, J.L., Davies, Z.G., McHugh, N., Gaston, K.J. & Leake, J.R. (2012) Organic carbon hidden in urban ecosystems. Scientific Reports, 2, 963.
22. Edmondson, J.L., Davies, Z.G., McCormack, S.A., Gaston, K.J. & Leake, J.R. (2014) Land-cover effects on soil organic carbon stocks in a European city. Science of the Total Environment, 472, 444–453.
23. ENA (2010) European Technical and Quality Standards for Nursery Stock. European Nurserystock Association, Lochristi, Belgium.
24. Esteban, B., Riba, J.-R., Baquero, G., Puig, R. & Rius, A. (2014) Environmental assessment of small-scale production of wood chips as a fuel for residential heating boilers. Renewable Energy, 62, 106–115.
25. Forestry Commission (2010) Tree Care Guide. Forestry Commission, London.
26. French, C.J., Dickinson, N.M. & Putwain, P.D. (2006) Woody biomass phytoremediation of contaminated brownfield land. Environmental Pollution, 141, 387–395.
27. Fuller, R.A., Irvine, K.N., Devine-Wright, P., Warren, P.H. & Gaston, K.J. (2007) Psychological benefits of greenspace increase with biodiversity. Biology Letters, 3, 390–394.
28. Gasson, P.E. & Cutler, D.F. (1998) Can we live with trees in our towns and cities? Arboricultural Journal, 22, 1–9.
29. Gilbertson, P. & Bradshaw, A.D. (1990) The survival of newly planted trees in inner cities. Arboricultural Journal, 14, 287–309.
30. Gómez-Baggethun, E., Gren, Å., Barton, D.N., Langemeyer, J., McPhearson, T., O'Farrell, P. (2013) Urban ecosystem services, Chapter 11. Urbanization, Biodiversity and Ecosystem Services: Challenges and Opportunities: A Global Assessment (eds T. Elmqvist et al.), pp. 175–251. SpringerOpen, Dordrecht.
31. Gruffydd, B. (1987) Tree Form, Size and Colour: A Guide to Selection, Planting and Design. E & FN Spon, London.
32. Hodge, S.J. & White, J.E.J. (1990) The Ultimate Size and Spread of Trees Commonly Grown in Towns. Arboriculture Research Note, Department of the Environment, London.
33. Holtsmark, B. (2013) The outcome is in the assumptions: analyzing the effects on atmospheric CO2 levels of increased use of bioenergy from forest biomass. GCB Bioenergy, 5, 467–473.
34. Infoterra (2006) LandBase Geoperspectives Layer Leicester (Updated November 2006). Infoterra, Leicester.
35. Kaplan, R. (2007) Employees' reactions to nearby nature at their workplace: the wild and the tame. Landscape and Urban Planning, 82, 17–24.
36. Kroeger, T., Escobedo, F.J., Hernandez, J.L., Varela, S., Delphin, S., Fisher, J.R.B. & Waldron, J. (2014) Reforestation as a novel abatement and compliance measure

for ground-level ozone. Proceedings of the National Academy of Sciences of the United States of America, 111, E4204–E4213.
37. LAEC (2007) Chainsaw Massacre: A Review of London's Street Trees. Greater London Authority, London.
38. Leicester City Council (2012) A Low Carbon City Climate Change – Leicester's Programme of Action. Leicester City Council, Leicester.
39. Loram, A., Tratalos, J., Warren, P.H. & Gaston, K.J. (2007) Urban domestic gardens (X): the extent and structure of the resource in five major cities. Landscape Ecology, 22, 601–615.
40. Luck, M. & Wu, J.G. (2002) A gradient analysis of urban landscape pattern: a case study from the Phoenix metropolitan region, Arizona, USA. Landscape Ecology, 17, 327–339.
41. MacFarlane, D.W. (2009) Potential availability of urban wood biomass in Michigan: implications for energy production, carbon sequestration and sustainable forest management in the U.S.A. Biomass and Bioenergy, 33, 628–634.
42. McHale, M.R., Burke, I.C., Lefsky, M.A., Peper, P.J. & McPherson, E.G. (2009) Urban forest biomass estimates: is it important to use allometric relationships developed specifically for urban trees? Urban Ecosystems, 12, 95–113.
43. McHugh, N., Edmondson, J.L., Gaston, K.J., Leake, J.R. & O'Sullivan, O.S. (2015) Modelling short-rotation coppice and tree planting for urban carbon management – a city-wide analysis. Dryad Digital Repository, http://dx.doi.org/10.5061/dryad.j25t0
44. McKinney, M.L. (2008) Effects of urbanization on species richness: a review of plants and animals. Urban Ecosystems, 11, 161–176.
45. McPherson, E.G. & Kendall, A. (2014) A life cycle carbon dioxide inventory of the Million Trees Los Angeles program. The International Journal of Life Cycle Assessment, 19, 1653–1665.
46. Met Office (2012) 1971–2000 Climate averages. Available at: http://www.metoffice.gov.uk/climate/uk/averages/19712000/ (accessed March 2014).
47. Milne, R. & Brown, T.A. (1997) Carbon in the vegetation and soils of Great Britain. Journal of Environmental Management, 49, 413–433.
48. Natural England (2013a) Opportunities and optimum sitings for energy crops. Available at: www.naturalengland.org.uk/ourwork/farming/funding/ecs/sitings/default.aspx (accessed March 2014).
49. Natural England (2013b) Energy Crops Scheme Establishment Grants Handbook. 3rd Edition Version 3.1. Natural England, Worcester.
50. Nielsen, A.B. & Møller, F.G. (2008) Is coppice a potential for urban forestry? The social perspective. Urban Forestry & Urban Greening, 7, 129–138.
51. Nowak, D.J. & Crane, D.E. (2002) Carbon storage and sequestration by urban trees in the USA. Environmental Pollution, 116, 381–389.
52. Nowak, D.J., Kuroda, M. & Crane, D.E. (2004) Tree mortality rates and tree population projections in Baltimore, Maryland, USA. Urban Forestry & Urban Greening, 2, 139–147.
53. Nowak, D.J., McBride, J.R. & Beatty, R.A. (1990) Newly planted street tree growth and mortality. Journal of Arboriculture, 16, 124–129.

54. Nowak, D., Hoehn, R., Bodine, A., Greenfield, E. & O'Neil-Dunne, J. (2013a) Urban forest structure, ecosystem services and change in Syracuse, NY. Urban Ecosystems, doi: 10.1007/s11252-013-0326-z.
55. Nowak, D.J., Greenfield, E.J., Hoehn, R.E. & Lapoint, E. (2013b) Carbon storage and sequestration by trees in urban and community areas of the United States. Environmental Pollution, 178, 229–236.
56. Ordnance Survey (2008) OS MasterMap topography layer. Updated June 2008.
57. Rahman, M., Armson, D. & Ennos, A. (2014) A comparison of the growth and cooling effectiveness of five commonly planted urban tree species. Urban Ecosystems, 17, 1–19.
58. Renewable Fuels Agency (2008) The Gallagher Review of the Indirect Effects of Biofuel Production. Renewable Fuels Agency, St Leonards-on-Sea.
59. RHS (2014) Trees. Available at: www.rhs.org.uk/plants/trees (accessed September 2014).
60. Rowe, R.L., Street, N.R. & Taylor, G. (2009) Identifying potential environmental impacts of large-scale deployment of dedicated bioenergy crops in the UK. Renewable and Sustainable Energy Reviews, 13, 271–290.
61. Roy, S., Byrne, J. & Pickering, C. (2012) A systematic quantitative review of urban tree benefits, costs, and assessment methods across cities in different climatic zones. Urban Forestry & Urban Greening, 11, 351–363.
62. Russo, A., Escobedo, F.J., Timilsina, N., Schmitt, A.O., Varela, S. & Zerbe, S. (2014) Assessing urban tree carbon storage and sequestration in Bolzano, Italy. International Journal of Biodiversity Science, Ecosystem Services & Management, 10, 54–70.
63. Sæbø, A., Popek, R., Nawrot, B., Hanslin, H.M., Gawronska, H. & Gawronski, S.W. (2012) Plant species differences in particulate matter accumulation on leaf surfaces. Science of the Total Environment, 427–428, 347–354.
64. St Clair, S., Hillier, J. & Smith, P. (2008) Estimating the pre-harvest greenhouse gas costs of energy crop production. Biomass and Bioenergy, 32, 442–452.
65. Stovin, V.R., Jorgensen, A. & Clayden, A. (2008) Street trees and stormwater management. Arboricultural Journal, 30, 297–310.
66. Strohbach, M., Arnold, E., Vollrodt, S. & Haase, D. (2012) Carbon sequestration in shrinking cities – potential or a drop in the ocean? Urban Environment (eds S. Rauch & G.M. Morrison), pp. 61–70. Springer, Dordrecht.
67. The Carbon Trust (2012) Biomass Heat Accelerator – Overview and Summary of Output. The Carbon Trust, London.
68. Timilsina, N., Staudhammer, C.L., Escobedo, F.J. & Lawrence, A. (2014) Tree biomass, wood waste yield, and carbon storage changes in an urban forest. Landscape and Urban Planning, 127, 18–27.
69. UK Parliament (2008) Climate Change Act c. 27. Available at: http://www.legislation.gov.uk/ukpga/2008/27 (accessed 14 February 2013).
70. UN-Habitat (2011) Cities and Climate Change: Global Report on Human Settlement 2011. United Nations Human Settlement Programme, London.
71. United Nations (2012) World Urbanization Prospects: The 2011 Revision. Department of Economic and Social Affairs, Population Division, New York, NY.

72. Willoughby, I., Stokes, V., Poole, J., White, J.E.J. & Hodge, S.J. (2007) The potential of 44 native and non-native tree species for woodland creation on a range of contrasting sites in lowland Britain. Forestry, 80, 531–553.
73. Wiström, B., Nielsen, A.B., Klobuĉar, B. & Klepec, U. (2015) Zoned selective coppice –a management system for graded forest edges. Urban Forestry & Urban Greening, 14, 156–162.
74. Wu, C., Xiao, Q. & McPherson, E.G. (2008) A method for locating potential tree-planting sites in urban areas: a case study of Los Angeles, USA. Urban Forestry & Urban Greening, 7, 65–76.
75. Zhao, X., Monnell, J.D., Niblick, B., Rovensky, C.D. & Landis, A.E. (2014) The viability of biofuel production on urban marginal land: an analysis of metal contaminants and energy balance for Pittsburgh's Sunflower Gardens. Landscape and Urban Planning, 124, 22–33.

Additional supplemental information (including tables, figures, and appendices) available online at http://onlinelibrary.wiley.com/doi/10.1111/1365-2664.12491/full.

CHAPTER 3

Neighborhood Greenspace and Health in a Large Urban Center

OMID KARDAN, PETER GOZDYRA, BRATISLAV MISIC, FAISAL MOOLA, LYLE J. PALMER, TOMÁŠ PAUS, AND MARC G. BERMAN

3.1 INTRODUCTION

Many have the intuition that living near trees and greenspace is beneficial to our health. But how much could a tree in the street or a nearby neighborhood park improve our health? Here we set out to examine this very question by studying the relationship between health and neighborhood greenspace as measured with comprehensive metrics of tree canopy on the street vs. tree canopy in parks and private residences.

It is a known fact that urban trees improve air quality[1,2], reduce cooling and heating energy use[3], and make urban environments aesthetically more preferable[4,5]. Importantly, several studies have shown that exposure to greenspaces can be psychologically and physiologically restorative by promoting mental health[6,7], reducing non-accidental mortality[8], reducing physician assessed-morbidity[9], reducing income-related health inequality's effect on morbidity[10], reducing blood pressure and stress levels[11,12], reducing sedentary leisure time[13], as well as promoting physical activity[14,15]. In addition, greenspace may enhance psychological and cardio-vascular benefits of physical activity, as compared with other settings[12].

Moreover, experimental research has demonstrated that interacting with natural environments can have beneficial effects—after brief exposures—on memory and attention for healthy individuals[16,17,18] and for patient populations[19,20,21]. In addition, having access to views of natural settings (e.g., from a home or a hospital bed) have been found to reduce crime and aggression[22,23] and improve recovery from surgery[24].

Although many studies have shown that natural environments enhance health or encourage healthy behaviors, to our knowledge, fewer studies have quantified the relationship between individual trees and health. In addition, studies have not separately estimated the treed area beside the streets and other urban greenspaces and related those variables to individuals' health in various domains, including cardio-metabolic conditions, mental disorders and general health perception. Knowing the kind of greenspace that may be associated with health benefits would be critical when deciding the type of greenspace that should be incorporated into built environments to improve health.

The typical method for quantifying exposure to greenspace for individuals in large population studies is to use the percentage of area covered in greenspace in an individual's neighborhood. The size of the areas and the accuracy (and also definition) of greenspace quantification vary across different studies. For example[10], used data containing $>10\,m^2$ accuracy for greenspace and geographical units of $4\,km^2$ on average in their study, Richardson et al. (2013) used >200 m2 accuracy for greenspace and geographical units that averaged $5\,km^2$, and7 used the presence of public "natural" spaces in areas within a 5 km radius from schools to quantify exposure to nature for school-aged children.

In this study, we were interested in examining greenspace with lower granularity (i.e., higher geographical resolution) and quantifying associations that are specific to exposure to trees, as opposed to exposures to any greenspace, such as grass or shrubbery. Here, our definition of greenspace consisted of tree canopy only and not of urban grass or bushes (or other "natural" settings). This choice is based on the assumption that trees are the most consistent green components in an area and potentially the most important component for having beneficial effects[25].

We also used a much higher geographical resolution for the following reasons. First, we wanted to distinguish between trees along the roads and streets versus those in domestic gardens and parks, and other open areas. To do so, we used individual tree data from the 'Street Tree General Data' and tree-canopy polygon data from the 'Forest and Land Cover' dataset to construct our greenspace variables. Both datasets came from the city of Toronto. Second, to ensure that the tree variables were less confounded by health insurance policies as well as demographic parameters (age, sex, education, and income), we used a single urban population (Toronto) in Canada, a country with a universal publically funded healthcare system that, compared with the United States, guarantees access to health-care services independent of income and/or employment status[26]. These health-care equalities facilitate the interpretation of the relationships between individual urban trees and health in this urban population. Although financial barriers may not impede access to health care services in Canada, differential use of physician services with respect to socio-economic status persist; Canadians with lower incomes and fewer years of schooling visit specialists at a lower rate than those with moderate or high incomes and higher levels of education despite the existence of universal health care[27]. In particular, we examined the relationship between tree canopy density beside the streets and in other areas such as parks and domestic gardens with an individual's health. The health variables that we focused on were: 1) Overall health perception; 2) Presence of cardio-metabolic conditions such as hypertension, high blood glucose, obesity (both overweight and obese), high cholesterol, myocardiac infarction, heart disease, stroke, and diabetes; and 3) Mental health problems including major depression, anxiety, and addiction. Subjective self-rated health perception was chosen as one of the health outcomes because self-perception of health has been found to be related to morbidity and mortality rates and is a strong predictor of health status and outcomes in both clinical and community settings[28,29,30].

Furthermore, on the tree variable side, we distinguished tree canopy of trees beside the street from those planted in other areas, such as parks and private backyards. A distinction of these different sources of tree canopy may be helpful for urban planning policies. We hypothesized that street trees could have stronger beneficial associations with individual's health because they may be more accessible to all residents in a given neighborhood as residents are likely exposed to street trees in their daily activities and through views from their windows; for example see[24].

Figure 1 shows a geographic map of the individual tree data (i.e., the individual trees on the street) and Fig. 2 shows a geographic map of the satellite tree data (i.e., the amount of tree canopy) for different neighborhoods in the city of Toronto. Both tree datasets were used to quantify the "greenness" of the neighborhoods (see Methods). Figure 3 shows the dissemination areas (i.e., Toronto neighborhood units) that were used in our analysis. The highlighted neighborhoods are the ones that were included in our analysis.

To uncover the relationships between neighborhood greenspace and health we performed two analyses. The first was a multiple regression of each health outcome on socio-economic, demographic and tree density variables. The second was a canonical correlation analysis where we examined the multivariate relationship between all health outcomes and socio-economic, demographic and tree density variables. Our canonical correlation model is shown in Fig. 4. In all of these analyses we attempted to quantify the independent relationships of street tree canopy and non-street tree canopy on health.

3.2 RESULTS

3.2.1 REGRESSION RESULTS

3.2.1.1 HEALTH PERCEPTION

Our results suggest that people who live in areas that have more (and/or larger) trees on the streets report better health perception, after controlling for demographic factors, such as income, age and education [$p<0.0001$]. As can be seen in Table 1, the regression coefficient for the street tree density variable shows that a four percent square meters

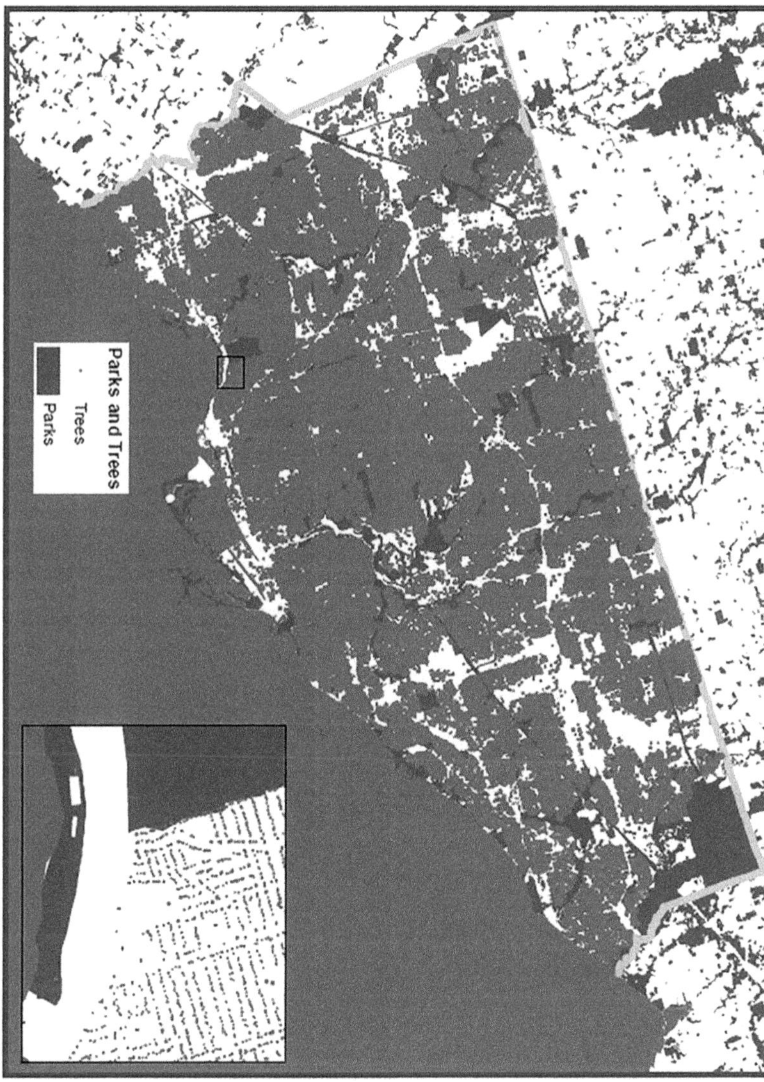

FIGURE 1: The Greenspace map of the city of Toronto constructed from the individual tree information Street Tree General Data. This image is shown in much lower resolution compared to the real image and the dissociation between individual trees and other areas is clearly perceivable for the zoomed-in area. Parks are shown in dark gray.

FIGURE 2: The Greenspace map of the city of Toronto constructed from the Geographical Information System (GIS) polygon data set Forest and Land Cover. The levels are shown in units of 10–15% for display purposes only as we analyzed these data as a continuous variable.

(400 cm^2) increase in the treed area for every square meter of neighborhood predicts about 0.04 increased health perception (i.e., 1% of our 1–5 health perception scale) for individuals living in that area. A 400 cm^2/m^2 increase in treed area is equal to the addition of about 200 average trees (with 40 m^2 crown area) on the streets in a dissemination area of almost average size (about 200,000 m^2) in Toronto. This is approximately 10 more trees per city block (a DA usually contains about 25 blocks). As can be seen in Table 1, this increase in health perception is equivalent to the effect of a $10,200 increase in annual household income and living in a DA with equally (i.e., $10,200) higher median income. (Notice that for this comparison we added up the estimates of income and area income because a hypothetical increase of income for the families in a DA also increases the median area income in that DA to the same extent). This same increase in health perception is also, on average, equivalent to being 7 years younger.

Other than street tree density, variables that independently predict better health perception in this multiple regression are: eating more servings of vegetables and fruits in one's diet (1 more serving per day predicts 1.2% better health perception [$p<0.0001$]), being younger (10 years less age predicts 1.5% better health perception [$p<0.0001$]), being male (males have on average almost 1% better health perception than females [$p=0.0004$]), having higher education (belonging to one higher educational group predicts 1.6% better health perception [$p<0.0001$]), living in more affluent neighborhoods (belonging to one higher area median income group predicts 0.7% better health perception [$p<0.0001$]), and having higher household income (belonging to one higher income group predicts 1.6% better health perception [$p<0.0001$]). It should be mentioned that the associations between health perception and tree density and other predictors reported here explain 9% of the variance in health perception. While the model explains a significant proportion of the variance in the data, it does not explain all of the variance of the dependent variable. This is true of all models whose R^2 values are less than 1. As such the model's predictions may not always hold true if the other unidentified factors that predict the remaining variability in health perception are not controlled for.

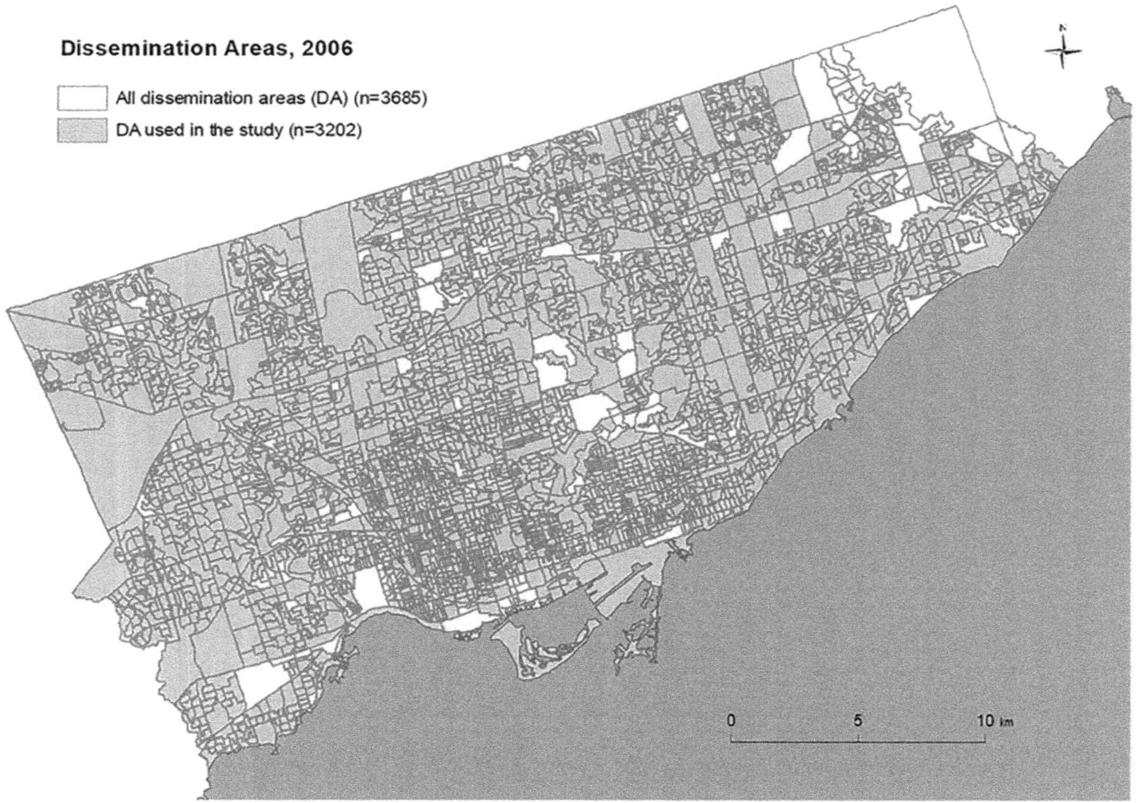

FIGURE 3: The dissemination area map of the city of Toronto (2006). The colored regions show the dissemination areas that were included in the study.

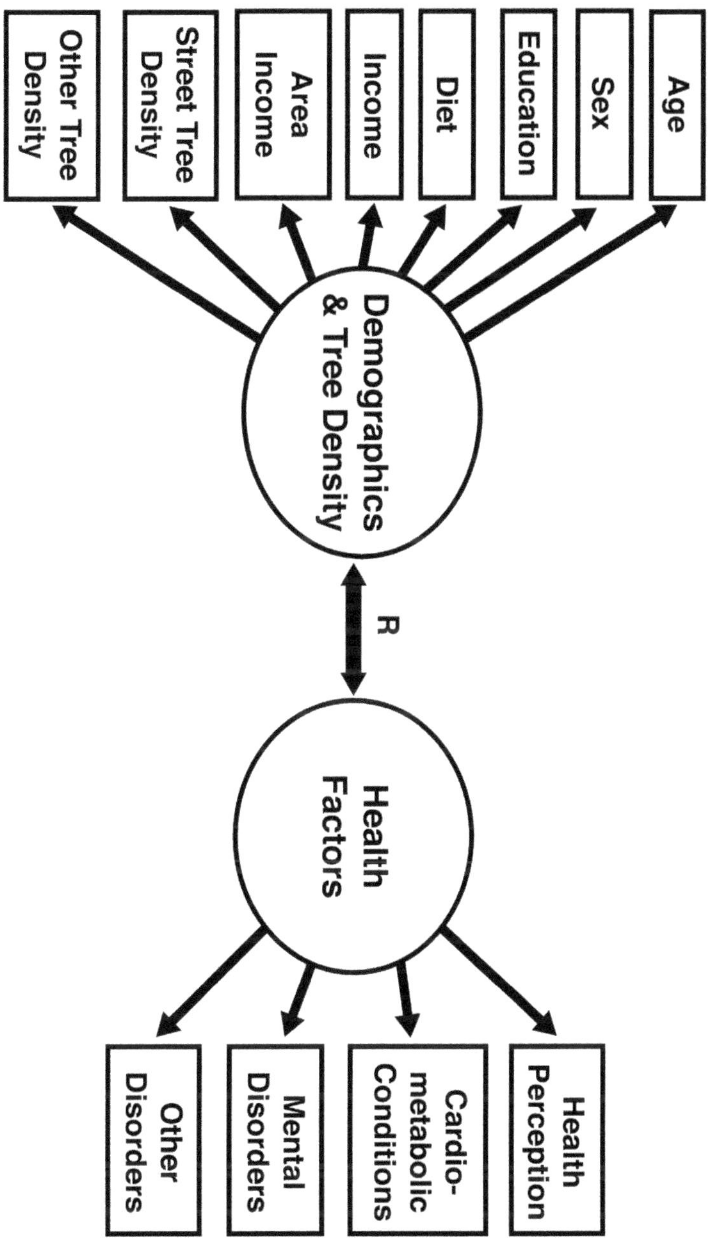

FIGURE 4: The canonical correspondence model that was used in our canonical correlation analyses to assess the relationship of the predictors (socio-economic, demographic and tree density variables) with health factors.

Table 1. Combined results of regression of health perception on the multiply-imputed data.

Variable	Estimate	Std. Error	t-stat	p-value	df	Rel. Increase	FMI
Intercept	2.7794	0.0296	93.8319	<0.0001	6202	0.0685	0.0644
Diet	0.0481	0.0024	19.7007	<0.0001	668	0.2130	0.1781
Age	–0.0059	0.0004	–16.8734	<0.0001	10566	0.05246	0.0500
Sex	0.0374	0.0107	3.4853	0.0004	14364	0.04498	0.0432
Education	0.0663	0.0032	20.6885	<0.0001	6647	0.06620	0.0624
Income	0.0710	0.0034	21.0145	<0.0001	448	0.2630	0.2117
Area income	0.0278	0.0056	4.9162	<0.0001	3664	0.08932	0.0825
Street Tree den.	0.0101	0.0015	6.6879	<0.0001	34158	0.02915	0.0284
Other Tree den.	–0.0003	0.0004	–0.7293	0.4658	25993	0.03342	0.0324

$R^2=0.0885$, adjusted $R^2=0.0876$, $F\ (8,\ 7879^*)=94.6814$, $p<0.0001$. FMI is fraction of missing information. *The average of estimated degrees of freedom.

3.2.1.2 CARDIO-METABOLIC CONDITIONS

Results of regressing the cardio-metabolic conditions index on the independent variables are shown in Table 2. Results suggest that people who live in areas that have more (and/or larger) trees on the streets report significantly fewer cardio-metabolic conditions. People reported decrease of 0.04 units of cardio-metabolic conditions (0.5% of the 0–8 scale for cardio-metabolic conditions) for every increase of 408 cm^2/m^2 in tree density. This is approximately equivalent to 11 more average-sized trees on the streets per city block. This effect for cardio-metabolic conditions is equivalent to a $20,200 increase in both area median income and annual household income adjusted for other variables. This decrease in cardio-metabolic conditions is also, on average, equivalent to being 1.4 years younger.

Other than street tree density, variables that predict fewer cardio-metabolic conditions, after controlling for other variables in this multiple regression, are: eating more servings of vegetables and fruits in one's diet (1 more serving per day predicts 0.08% less cardio-metabolic conditions [$p=0.0129$]), being younger (10 years less age predicts 3.7% less cardio-metabolic conditions [$p<0.0001$]), being female (females report on average 3.3% less cardio-metabolic conditions than males [$p<0.0001$]), having higher education (belonging to one higher educational group predicts 0.71% less cardio-metabolic conditions [$p<0.0001$]), living in more affluent neighborhoods (belonging to one higher area median income group predicts 0.36% higher reported health perception [$p<0.0001$]), and having higher household income (belonging to one higher income group predicts 0.28% less cardio-metabolic conditions [$p<0.0001$]). In addition, we added the interaction terms of all predictors with the tree density variables and the models R^2 for health perception and cardio-metabolic conditions did not improve much ($\Delta R^2=0.0008$ for health perception, $\Delta R^2=0.0009$ for cardio-metabolic conditions), even though there was a small positive interaction between street tree density and age that was statistically significant. We chose not to include these interactions due to lack of a priori hypotheses, their small effect sizes and to preserve the models simplicity. Again, it should be mentioned that the associations between cardio-metabolic conditions and tree density and other predictors reported here explain 19% of the variance in cardio-metabolic conditions. While the model explains

Table 2. Combined results of regression of cardio-metabolic conditions on the multiple-imputed data.

Variable	Estimate	Std. Error	t-stat	p-value	df	Rel. Increase	FMI
Intercept	0.1236	0.0363	3.4049	0.0008	895	0.1937	0.1643
Diet	–0.0062	0.0026	–2.3217	0.0204	1206	0.1569	0.1371
Age	0.0296	0.0004	70.4279	<0.0001	1724	0.1307	0.1166
Sex	0.2894	0.0128	22.5830	<0.0001	857	0.1871	0.1596
Education	–0.0570	0.0037	–15.2098	<0.0001	553	0.2351	0.1932
Income	–0.0240	0.0038	–6.2648	<0.0001	168	0.4563	0.3213
Area income	–0.0286	0.0066	–4.3071	<0.0001	863	0.1864	0.1591
Street Tree den.	–0.0097	0.0018	–5.4025	<0.0001	801	0.1937	0.1643
Other Tree den.	–0.0001	0.0005	–0.1196	0.9048	776	0.1970	0.1667

$R^2 = 0.1920$, adjusted $R^2 = 0.1845$, $F(8, 871^*) = 25.6089$, $p < 0.0001$. FMI is fraction of missing information. *The average of estimated degrees of freedoms.

a significant proportion of the variance in the data, it does not explain all of the variance of the dependent variable. This is true of all models whose R^2 values are less than 1. As such the model's predictions may not always hold true if the other unidentified factors that predict the remaining variability in cardio-metabolic conditions are not controlled for.

3.2.1.3 MENTAL DISORDERS AND OTHER DISORDERS

Results of Mental Disorders and Other Disorders can be found in Supplemental Tables S1 and S2. Regressing the Mental Disorders index on the independent variables do not capture a significant amount of variance in Mental Disorders in the data [$R^2=0.0136$, adjusted $R^2=-0.0111$, $p=0.1820$]. We will further investigate this issue later in the canonical correlation analysis.

Finally, the Other Disorders index is not a coherent variable and was not constructed to be used as a dependent variable in the regression analyses, but mainly was constructed as a control variable for the canonical correlation analysis. Nonetheless, results of regressing the Other Disorders index (Cancer, Migraines, Arthritis, or Asthma) on the independent variables are shown in Table S2.

3.2.2 CANONICAL CORRELATION RESULTS

Figures 5, 6, 7 show the results from the canonical correlation analysis, which finds the relationship (i.e., linear combination of weights) between two sets of variables. The height of each bar shows the correlation of each variable with the corresponding set of canonical weights. Error bars show ±2 standard errors containing both between and within imputation variance calculated by bootstrapping imputed data sets. Importantly, all canonical variates are orthogonal to one another.

The canonical correlation coefficient (r) for each pair of linear composites is shown near the bidirectional arrow representing the relationship between the two sets of variables (demographic and green-space variables and health-related variables). The canonical correlation coefficients for all the four pairs of linear composites were statistically significant ($p<0.0001$ for Bartlett's approximate chi-squared statistic with Lawley's modification).

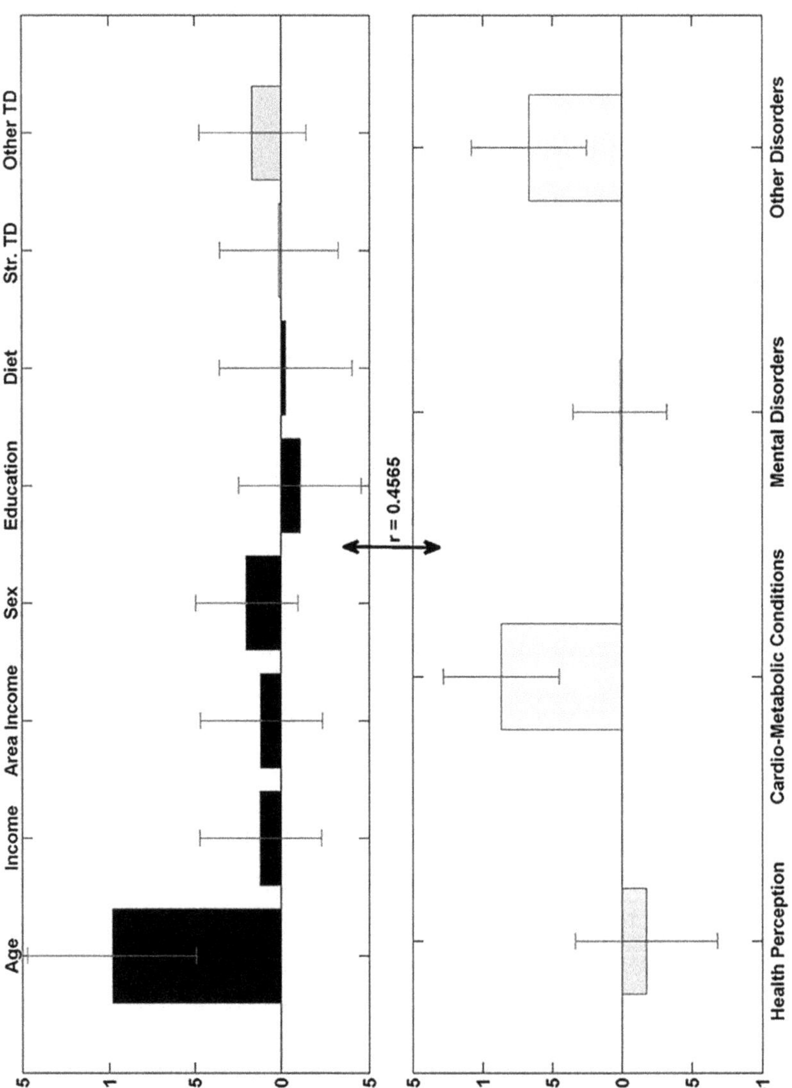

FIGURE 5: The first pair of linear composites for the canonical correlation analysis; $F (32, 114680)=381.2263)$, $R^2=0.2084$, $p<0.0001$. Bars show correlation of each variable (canonical loadings) with the first set of weighted canonical scores. Error bars show ±2 standard errors containing both between and within imputation variance calculated by bootstrapping imputed data sets.

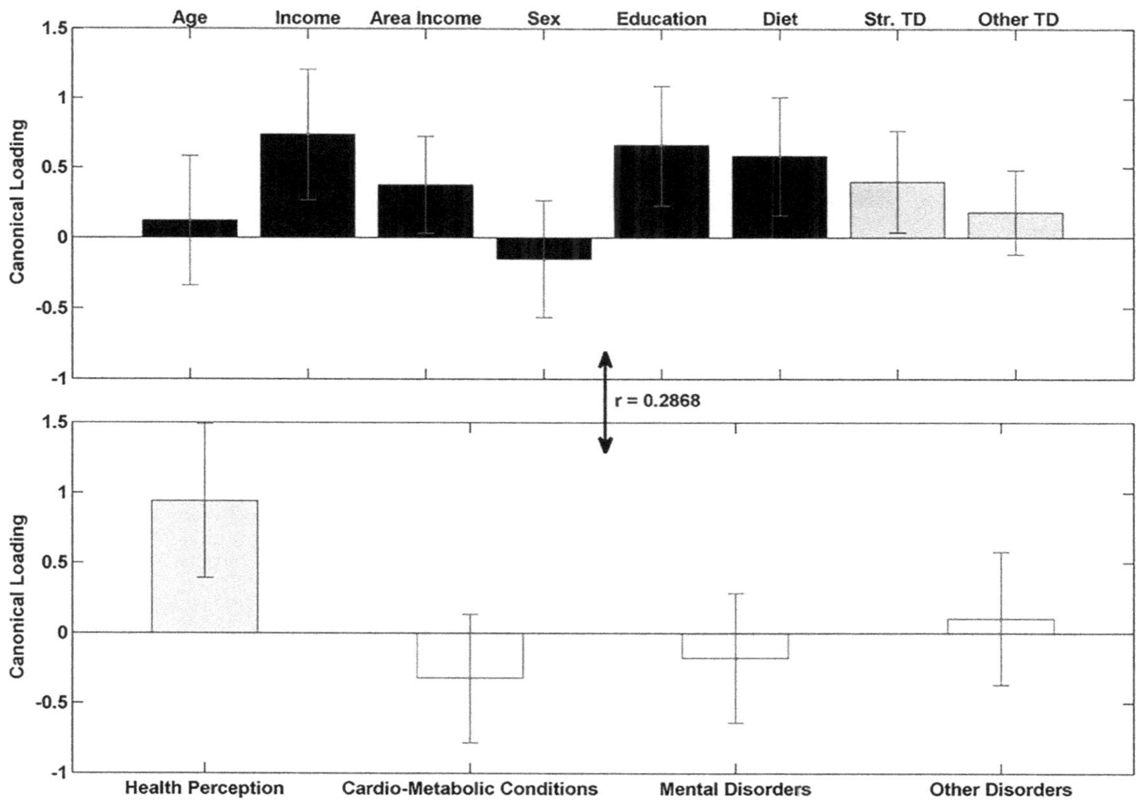

FIGURE 6: The second pair of linear composites for the canonical correlation analysis; F (21, 89297) = 211.0480), R^2 = 0.0822, $p < 0.0001$. Bars show correlation of each variable with the second set of weighted canonical scores. Error bars show ±2 standard errors containing both between and within imputation.

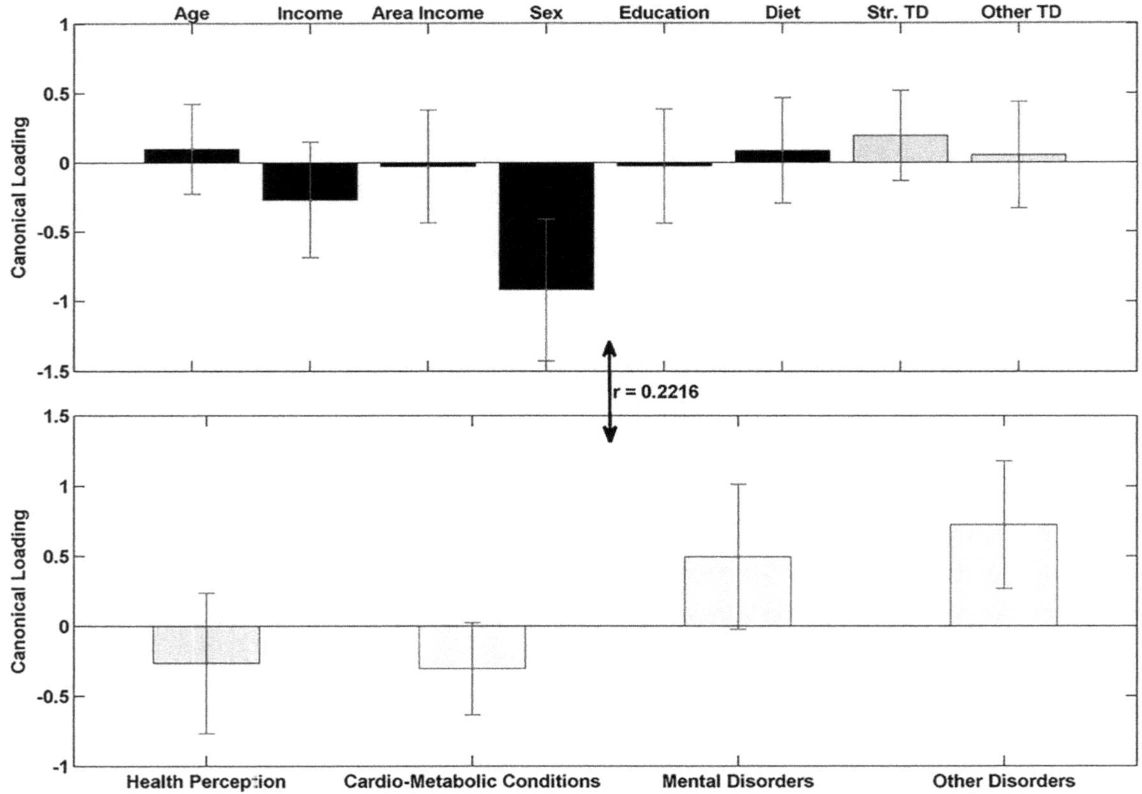

FIGURE 7: The third pair of linear composites for the canonical correlation analysis; F (12, 63702)=139.9347, R^2=0.0491, p<0.0001. Bars show correlation of each variable with the third set of weighted canonical scores. Error bars show ±2 standard errors containing both between and within imputation variance.

The first pair of linear composites (Fig. 5) is dominated by the effect of age on physical disorders (Cardio-metabolic and Other disorders). This suggests that being older is highly correlated ($r=0.4565$, $R^2=0.2084$) with having more cardio-metabolic conditions, as well as cancer, arthritis, asthma and migraines.

The second pair of linear composites is mainly dominated by Health Perception and shows that individuals with higher annual income, higher education, higher vegetables/fruits consumption and who live in areas with higher street tree density report the best health perception. This replicates and extends the results found in the regression. The same group of people also reports fewer cardio-metabolic conditions, although the errorbar for the loading of these conditions crosses zero (indicating a non-significant effect). This is possibly due to the fact that the main part of the variability in cardio-metabolic conditions (that was mainly due to older age) was already captured by the first canonical loadings. The canonical correlation for this second linear composite is of medium size ($r=0.2868$, $R^2=0.0822$).

The third pair of linear composites has a modest effect size ($r=0.2216$, $R^2=0.0491$) and is mainly dominated by sex. This composite shows that females report more other disorders and more mental disorders. This complies with the regression results and the fact that occurrence of breast cancer is more frequent among women even at younger ages[31].

Results from the fourth composite are shown in Supplementary Figure S1. The fourth component was dominated by mental disorders after much of the variability due to sex was extracted by the previous composites (mainly third composite). Neither the demographic nor the tree density variables significantly correlated with the fourth canonical scores. The very small effect size ($r=0.0539$, $R^2=0.0029$) shows that the data and variables might not be rich enough for an analysis of mental disorders, as mentioned before in the regression analysis. Indeed, only a non-reliable combination of demographic and tree variables seem to be related to more mental disorders at this stage of analysis. Future studies with more detailed data regarding mental disorders may help to test the results found for the fourth composite.

Finally, Table 3 shows the communalities for all the variables, which are computed as sum of the squared loadings across all latent variables and

Table 3. Communalities for the variables based on the canonical correlation analysis.

Variable	Communal-ity	Variable	Communality
Age	0.9845	Str. Tree Density	0.3980
Income	0.8158	Other Tree Density	0.1317
Area Income	0.2649	Health Perception	1.0000
Sex	0.9848	Cardio-metabolic Conditions	1.0000
Education	0.5016	Mental Disorders	1.0000
Diet	0.4372	Other Disorders	1.0000

represent how much of the variance in the variable has been accounted for by the canonical correlation model. The communality results show that the canonical variates are able to capture/reproduce at least 15% of the variance in all original variables. In conclusion, both the regression and the canonical correlation analyses suggest that higher tree density on the streets, in a given dissemination area, correlates with better health perception and fewer cardio-metabolic conditions for people living in that dissemination area.

3.3 DISCUSSION

Results from our study suggest that people who live in areas with higher street tree density report better health perception and fewer cardio-metabolic conditions compared with their peers living in areas with lower street tree density. There are two important points about our results that add to the previous literature. First, the effect size of the impact of street tree density seems to be comparable to that of a number of socioeconomic or demographic variables known to correlate with better health (beyond age). Specifically, if we consider two families, one earning $10,200 more annually than the other, and living in a neighborhood with the same higher median income, it is predicted that the more affluent family who is living in the richer neighborhood perceives themselves as healthier people. Interestingly, however, that prediction could turn out to be wrong if the less affluent family lives in a neighborhood that has on average 10 more trees beside the streets in every block. Regarding cardio-metabolic conditions, the same scenario is expected to hold true for an income difference of $20,200.

Ten more trees in every block is about 4% increase in street tree density in a dissemination area in Toronto, which seems to be logistically feasible; Toronto's dissemination areas have a 0.2% to 20.5% range of street tree density and trees can be incorporated into various planting areas along local roads. According to our findings improving health perception and decreasing cardio-metabolic conditions by planting 10 more trees per city block is equivalent to increasing the income of every household in that city block by more than $10,000, which is more costly than planting the additional 10 trees. (See the "Urban Watershed Forestry Manual, Part 3 Urban Tree Planting Guide" for estimation of urban tree planting and maintenance costs and other considerations for urban tree planting. Generally, planting and maintenance of 10 urban trees could annually cost between $300 to $5000). Finally, it should be mentioned that this estimation of increased tree density being equivalent to specific increases in economic status of people is based on respondents from Canada, which has a publically funded universal health-care system. It may be the case that in other countries that do not have universal health care individuals' health may be more affected by economic status, which could cause the tree density relationship with health to be smaller-in economic terms. This, however, is an empirical question that is certainly worthy of further investigation.

The second important finding is that the "health" associations with tree density were not found (in a statistically reliable manner) for tree density in areas other than beside the streets and along local roads. It seems that trees that affect people most generally are those that they may have the most contact (visual or presence) with, which we are hypothesizing to be those planted along the streets. Another possible explanation could be that trees on the street may be more important to reductions in air pollution generated by traffic through dry deposition[32]. This does not indicate, however, that parks are not beneficial. This study only shows that planting trees along the roads may be more beneficial than planting trees in parks and private residences at least for these health measures. For example, our sample only consists of adults and trees in parks may be more beneficial to children who spend more time in such locations[33]. Future studies need to address this possibility more thoroughly.

An important issue that is not addressed in this study is the mechanisms by which these beneficial effects of proximity to more (or larger) urban trees on health occur. Improving air quality, relieving stress, or promoting physical activity could all be contributing factors to improved reported health. The current study provides two pieces of information that could be useful when trying to study the underlying mechanisms of the health benefits attained from urban trees. First, more than proximity (tree density in the neighborhood), it is the availability of the trees to the largest proportion of people (trees on the roads) that is beneficial. Second, the form of the relationship is linear, at least in the density range of 0 to 20% for trees on the streets found in the city of Toronto (i.e., adding the quadratic or the square root of street tree density to the multiple regressions did not improve the models, suggesting that the relationship of health outcomes with street tree density neither decreases (quadratic transformation), nor increases (square root transformation) in a meaningful way at higher levels of street tree density). These two results imply that: 1) some of the effects may be partially related to the mere visual exposure to trees[16,18,24] or to the dry deposition of air pollutants and 2) that the effects are not likely to plateau or accelerate, in a meaningful way, as the level of tree canopy density increases.

In addition, in a post-hoc analysis, we compared the health outcomes of individuals living in areas with more leaf-retaining versus more deciduous trees, adjusted for street and other tree density and demographic variables. Our analysis showed that people living in year-round green areas (more leaf-retaining trees) reported less cardio-metabolic conditions ($p=0.017$) than their peers, but not better health perception. Again, while not conclusive, this result points to some importance regarding the types of trees that should be planted, but it would be much too premature to favor the planting of non-deciduous vs. deciduous trees.

Our study could benefit from improvements in at least three aspects. First, we used cross-sectional data for practical reasons; longitudinal data would provide us with much stronger inferences of causality. Second, our health data items are self-reported, which introduces some error and potential biases in health variables reported. Third, we are assuming that controlling for area median income accounts for many other neighborhood variables that could affect mental and physical health in indirect ways (such as neighborhood safety, pollution, etc.), which might not always hold true. In future research we plan to test our current findings in a more comprehensive manner that obviates the mentioned limitations. In summary, our results show that street trees are associated with a significant, independent and reliable increase in health benefits in urban populations and that small increases in the number of trees along the street could improve health markedly and in cost-effective ways.

3.4 MATERIALS AND METHODS

Canada is divided into geographical units called dissemination areas (DA), which consist of 400 to 700 inhabitants and whose boundary lines mostly follow roads. We used data from 3,202 DAs located in the city of Toronto with an average population of 690 individuals and average physical size of 172,290 m^2.

We combined data from three different sources to construct our tree, health and demographic variables:

The first source of tree canopy data came from the 'Street Tree General Data,' which is a Geographical Information System (GIS) dataset that lists the locations of over 530,000 individual trees planted on public land within the city of Toronto. This dataset comes from experts who traversed the city of Toronto and recorded tree species and diameters at breast height. Trees in public parks are not listed as the listed trees were only from public land that lines the streets. The set contains each tree's common and botanical names, their diameters at breast height (DBH), the street addresses and the general location reference information. Figure 1 shows the green-space map of Toronto generated from these data for illustration.

The second source of tree canopy data came from the Geographical Information System (GIS) polygon data set 'Forest and Land Cover,' which contained detailed areal information of tree canopies in Toronto. In these data, the satellite imagery resolution was 0.6 m—from QuickBird Satellite imagery, 2007. The treed area was calculated using automated remote sensing software—Ecognition. This automated land-cover map was then monitored by staff from the University of Vermont Spatial Analysis Lab and adjusted to increase accuracy. In this dataset there is the ability to differentiate shrub cover from trees. There is, however, some susceptibility to errors when differentiating large shrubs from trees. To validate the accuracy of the QuickBird satellite imagery, it was compared with two other methods used to assess tree canopy cover: 1) Ocular estimates of canopy cover by field crews during data collection in 2008; 2) 10,000 random point samples of leaf-off and leaf-on aerial orthophotos (imagery available in required orthorecitifed format included 1999, 2005 and 2009)[34]. The tree canopy coverage estimates for each of the respective approaches were: QuickBird: 28%; Ocular: 24%; and Aerial Orthophotos: 26.2% respectively[34]. Because of the similarity in results, we can be confident in the accuracy of the QuickBird satellite results. For more information on the automated classification of leaf-on tree canopy from the 2007 satellite imagery see Appendix 4 of[34]. Figure 2 shows a map of tree canopy in each dissemination area as generated from the QuickBird Satellite.

Information about individuals' health and demographics was obtained in the context of the Ontario Health Study (https://www.ontariohealthstudy.ca). This is an ongoing research study of adults (18 years and older) living in the Canadian province of Ontario aimed at investigating risk factors associated with diseases such as: cancer, diabetes, heart disease, asthma, and Alzheimer's Disease. The data were collected using two (similar) versions of a web-based questionnaire consisting of demographic and health-related questions. These questionnaires were completed by 94,427 residents living in the greater Toronto area between September, 2010 and January, 2013. For this study, we used data from a subset of 31,109 residents (31,945 respondents, out of which 827 were removed during quality control for having duplicate records and 9 were removed because of missing consent records). A record was considered a duplicate with the following data quality checks: 1) Multiple registrations of the same Last Name, First Name and Date of Birth 2) Multiple registrations of the same Last Name, First Name and Postal Code 3) Multiple registrations of the same Last Name, First Name, Date of Birth and Postal Code 4) Multiple registrations of the same email address. Additional data quality checks included several built-in checks in the online system, which included automatic skip patterns and limited ranges for free text numerical responses such that participant responses must be within reasonable limits. The final sample included individuals who resided in the 3,202 dissemination areas of the city of Toronto as individual tree data were only available for these areas. These dissemination areas are shown in Fig. 3.

3.4.1 DEMOGRAPHIC VARIABLES

For each individual, we used sex (59% female; compared to the population male/female ratio: Toronto's population was 48.0% male and 52.0% female in 2011 according to Statistics Canada), age (Mean=43.8, range=18–99; as of 2011 the mean age of residents above 19 years of age for the entire population of Toronto is: 47.9 according to Statistics Canada), education (coded as: 1=none (0.0%), 2=elementary (1.0%), 3=high school (15.3%), 4=trade (3.3%), 5=diploma (15.9%), 6=certificate (5.9%), 7=bachelor's (35.3%), 8=graduate degree (23.3%), with Mean=6.07, range=1–8; According to the 2011 National Household Survey in www.toronto.ca, the distribution of education for the entire city of Toronto is the following: 33% of all City residents 15 years and over have a bachelor degree or higher, 69% of City residents between the ages of 25 and 64 years have a postsecondary degree, 17% of 25–64 years old residents have graduate degrees), and annual household income (coded as: 1=less than $10 000, 2=$10 000–$24 999, 3=$25 000–$49 999, 4=$50 000–$74 999, 5=$75 000–$99 999, 6=$100 000–$149 999, 7=$150 000–$199 999, 8=$200 000 or more, with Mean=4.67 which is equivalent to $90 806 annual income range=1–8; compared to the entire city of Toronto's population mean household income, which was: $87,038 in 2010 according to Statistics Canada), as well as diet (number of fruits and vegetable servings respondent consume every day, with Mean=2.24, range=0–10), as potential confounding variables. In addition, for each dissemination area we used the area median income from Statistics Canada and coded those data the same as the household income data, with mean=4.08, range=2–8. Population densities in a given DA were used in the multiple imputation analysis but not as a variable in the regressions or the canonical correlation analyses. The correlations between demographic variables can be found in Figure S2 of Supplementary Information.

Our studied sample had similar demographics to the entire city of Toronto, but was slightly younger (mean age=43.8; Toronto population=47.9), slightly more female (59%; Toronto population=52%), slightly more educated (35.3% had bachelor's degrees vs. 33% in the Toronto population) and slightly wealthier (mean household income=$93,399 vs. $87,038 in the entire city of Toronto).

3.4.2 GREEN-SPACE VARIABLES

Crown area of the trees was used to calculate the density of area covered by trees separately for the trees on the streets and the trees from greenspace in private locations and parks in each DA. We estimated the crown area of the trees based on their diameter at breast height (DBH) values. We obtained formulas for estimating tree crown diameter based on DBH for 8 tree types (Maple, Locust, Spruce, Ash, Linden, Oak, Cherry, and Birch) that were derived from forestry research. Forestry researchers have fit linear and non-linear models to relate crown diameter and DBH for different species of trees. These models achieved good fits as verified by their high R2 values (above 0.9)35,36. The formulas that were used to estimate crown diameters from DBH for these tree types and their references can be found in the Supplementary Equations section of the Supplementary Information. These 8 tree species covered 396,121 (83%) of the trees in our dataset. For the other 81,017 (17%) of the trees, we estimated crown diameter based on the linear regression of crown diameters on DBHs obtained from the 83% of the trees belonging to the tree types with known crown formulas. The crown areas of all the trees were then calculated using the crown diameters and assuming that the crown areas were circular in shape.

Street tree density for each dissemination area was quantified as the total area of the crowns of trees (m^2) beside the streets in the dissemination area over total dissemination area size (m^2) multiplied by 100 to be in percentage format. The range for this variable was found to be from 0.02% in the areas with the least street tree density to 20.5% in the areas with highest street tree density (Mean=4.57%). Other tree density for each dissemination area was calculated by subtracting out the area covered by crowns of the trees on the streets (street tree area) from the total treed area (m^2) in the dissemination area (from the satellite Tree Canopy data), and then dividing that by the area size and multiplying by 100 to be in percentage format. The range for this variable was found to be from 0.00% in the areas with almost no trees in parks (or no parks), no domestic gardens or other open areas; to 75.4% in areas with high tree density and parks (Mean=23.5%). As mentioned above, there was limited ability to differentiate large shrub cover from tree cover in the satellite data. Therefore, the variable "other tree density" could contain some unwanted large shrub cover as well, especially for areas with very high other tree density.

3.4.3 HEALTH VARIABLES

All of the health variables were constructed from the self-reported items in the Ontario Health Study (OHS). Items related to disorders were based on the question "Have you ever been diagnosed with …?" and coded with 0=No and 1=Yes. These consisted of physical conditions including high blood pressure, high cholesterol, high blood glucose, heart attack (MI), stroke, heart disease, migraines, chronic obstructive pulmonary disorder (COPD), liver cirrhosis, ulcerative colitis, irritable bowel disease (IBD), arthritis, asthma, cancer, and diabetes (DM), as well as mental health conditions including addiction, depression, and anxiety. About 66.3% of all respondents reported having at least one of the mentioned health conditions. The percentages of "Yes" responses for each of these conditions are reported in Supplementary Table S3. Additionally, body mass index (BMI) for each person was calculated from his/her self-reported height and weight. Our "Obesity" variable was constructed as 0 for BMI below 25, 0.5 for BMI between 25 and 30 (overweight, 26% of respondents), and 1 for BMI over 30 (obese, 13% of respondents). Other variables drawn from these data are general health perception (self-rated health (1=poor, 2=fair, 3=good, 4=very good, 5=excellent, with Mean=3.66, range=1–5), and four more variables that were used in the multiple imputations to increase the accuracy of imputations: walking (the number of days a participant has gone for a walk of at least 10 minutes in length last week, with Mean = 5.33, range = 0–7), smoking (if participant has ever smoked 4-5 packs of cigarettes in their lifetime, 38% Yes), alcohol consumption frequency (coded as 0 = never, 1 = less than monthly, 2 = about once a month, 3 = two to three times a month, 4 = once a week, 5 = two to three times a week, 6 = four to five times a week, with Mean = 3.60, range = 0–7), and alcohol binge frequency (coded as 0 = never, 1 = 1 to 5 times a year, 2 = 6 to 11 times a year, 3 = about once a month, 4 = 2 to 3 times a month, 5 = once a week, 6 = 2 to 3 times a week, 7 = 4 to 5 times a week, 8 = 6 to 7 times a week, with Mean = 1.62, range = 0–8).

The dependent variables related to physical and mental health were created from the multiple-imputed data. For each complete dataset, the Cardio-metabolic Conditions index was constructed by summing the following seven variables related to cardio-metabolic health: High Blood Glucose, Diabetes, Hypertension, High Cholesterol, Myocardial infarction (heart attack), Heart disease, Stroke, and "Obesity" with Mean = 0.89, range = 0–8. The Mental disorders index was constructed by summing Major Depression, Anxiety, and Addiction, with Mean = 0.26, range = 0–3. The Health Perception index was the third dependent variable in our analyses with Mean = 3.66, range = 1–5. The Other disorders index consisted of Cancer, Migraines, Asthma, and Arthritis (Mean = 0.48, range = 0–4. This index was constructed to be a control variable in the canonical correlation analysis. The additional variables (e.g., cirrhosis) were included to increase the accuracy of the imputation, but were not analyzed. The correlation matrix between the health variables, the tree variables, and the demographic variables is reported in supplementary Figure S2 of the Supplementary Information.

3.4.4 MULTIPLE IMPUTATIONS ANALYSIS

The self-reported health data contained some missing values for different variables (mainly due to "I don't know" responses). List wise deletion of the data (keeping only participants with no missing values in any of the items) would have resulted in a loss of 73% of the participants because the missing values in the different items were distributed across subjects, and was therefore an unreasonable method of analysis. To handle the missing data problem, we assumed that the data were missing at random (MAR), meaning that the probability of missingness for a variable was not dependent on the variable's value after controlling for other observed variables. We then replaced the missing values with multiple imputed data[37,38,39]. Thirty complete datasets were created from the original dataset using the estimate and maximize (EM) algorithm on bootstrapped data implemented by the Amelia package for R [Amelia[40];]. All of the 30 imputations converged in less than 11 iterations. Variables used in the imputations and their missing percentages are reported in Supplementary Table S4.

3.4.5 REGRESSION ANALYSIS

The regression analyses were performed separately for each imputed dataset and then combined based on Rubin's rules[38] using the Zelig program in R41. Rubin suggested that the mean of each regression coefficient across all imputed datasets be used as the regression coefficients for the analysis. In addition, to avoid underestimation of standard errors and taking the uncertainty of the imputed values into account, both the within imputation variance and between imputation variance of each coefficient should be used to construct the standard error for each regression coefficient. Lastly[42], proposed using degrees of freedom estimated as a function of the within and between imputation variance and the number of multiple imputations when approximating the t-statistics for each parameter.

To assess the amount of the variance in the dependent variables that is explained by the regression model for the multiple imputed data we used the method suggested by Harel (2009) to estimate the R^2 and the adjusted R2 values. Based on this method, instead of averaging R^2 values from the 30 imputations, first the square root of the R^2 value (r) in each of the imputed datasets is transformed to a z-score using Fisher's r to z transformation, z=atanh(r). The average z across the imputations can then be calculated. Finally, the mean of the z values is transformed back into an R^2. The same procedure can be used for adjusted R^2 values. Harel (2009) suggests that the number of imputations and the sample size be large when using this method, which holds true in the current study. Also, the resulting estimates of R^2 could be inflated (i.e. are too large), while estimates of adjusted R^2 tend to be biased downwards (i.e. are too small). Therefore, we estimated both values for a better evaluation of the explained variance.

3.4.6 CANONICAL CORRELATION ANALYSIS

To investigate further the relationship between the two sets of variables, namely the health-related variables (Health Perception, Cardio-metabolic conditions, Mental Disorders, and Other Disorders) and the demographic and green-space variables (Age, Sex, Education, Income, Area income, Diet, Street Tree Density, and Other Tree Density), we performed a canonical correlation analysis[43,44]. Our model is presented in the diagram shown in Fig. 4. Mauchly's test of sphericity was performed on the average of imputations in MATLAB (Sphertest: Sphericity tests menu) and showed that the correlation matrix of the data is significantly different from the identity matrix ($p<0.0001$). This significant departure of the data from sphericity warrants the canonical correlation analysis.

In a canonical correlation analysis, first, the weights that maximize the correlation of the two weighted sums (linear composites) of each set of variables (called canonical roots) are calculated. Then the first root is extracted and the weights that produce the second largest correlation between sum scores is calculated, subject to the constraint that the next set of sum scores is orthogonal to the previous one. Each successive root will explain a unique additional proportion of variability in the two sets of variables. There can be as many canonical roots as the minimum number of variables in the two sets, which is four in this analysis. Therefore, we obtain four sets of canonical weights for each set of variables, and each of these four canonical roots have a canonical correlation coefficient which is the square root of the explained variability between the two weighted sums (canonical roots).

To obtain unbiased canonical weights for variables and canonical correlation coefficients, we averaged data values over the 30 imputations and performed canonical correlation analysis on the z-scores of the averaged data using MATLAB (MATLAB and Statistics Toolbox Release 2014a, The MathWorks, Inc., Natick, Massachusetts, United States). For a more straight-forward interpretation and better characterization of the underlying latent variable, instead of using the canonical weights, we calculated the Pearson correlation coefficient (canonical loading) of each observed variable in the set with the weighted sum scores for each of the four linear composites. This way, each canonical root (linear composite) could be interpreted as an underlying latent variable whose degree of relationship with each of the observed variables in the set (how much the observed variable contributes to the canonical variate) is represented by the loading of the observed variable and its errorbar (see canonical correlation results).

To estimate the standard errors of the canonical loadings, we bootstrapped z-scores from each of the 30 complete imputed data (1000 simulations for each) and performed canonical correlation analysis 30000 times using MATLAB. Then, we calculated the variances of the set of loadings, which were calculated as explained above, over each completed dataset (within imputation variance). We also calculated the variance of the 30 sets of coefficients (between imputation variance). The standard errors of the coefficients were then estimated using the same Rubin's rules as was done for the regression analyses.

REFERENCES

1. Nowak, D. J., Crane, D. E. & Stevens, J. C. Air pollution removal by urban trees and shrubs in the United States. Urban forestry & urban greening 4, 115–123 (2006).
2. Nowak, D. J., Hirabayashi, S., Bodine, A. & Greenfield, E. Tree and forest effects on air quality and human health in the United States. Environmental Pollution 193, 119–129 (2014).
3. Akbari, H., Pomerantz, M. & Taha, H. Cool surfaces and shade trees to reduce energy use and improve air quality in urban areas. Solar energy 70, 295–310 (2001).
4. Kaplan, S., Kaplan, R. & Wendt, J. S. Rated preference and complexity for natural and urban visual material. Perception & Psychophysics 12, 354-&, 10.3758/bf03207221 (1972).
5. Smardon, R. C. Perception and aesthetics of the urban-environment - review of the role of vegetation. Landscape and Urban Planning 15, 85–106, 10.1016/0169-2046(88)90018-7 (1988).
6. Richardson, E. A., Pearce, J., Mitchell, R. & Kingham, S. Role of physical activity in the relationship between urban green space and health. Public health 127, 318–324 (2013).

7. Huynh, Q., Craig, W., Janssen, I. & Pickett, W. Exposure to public natural space as a protective factor for emotional well-being among young people in Canada. BMC public health 13, 407 (2013).
8. Villeneuve, P. J. et al. A cohort study relating urban green space with mortality in Ontario, Canada. Environmental Research 115, 51–58, 10.1016/j.envres.2012.03.003 (2012).
9. Maas, J. et al. Morbidity is related to a green living environment. Journal of epidemiology and community health 63, 967–973 (2009).
10. Mitchell, R. & Popham, F. Effect of exposure to natural environment on health inequalities: an observational population study. Lancet 372, 1655–1660, 10.1016/s0140-6736(08)61689-x (2008).
11. Grahn, P. & Stigsdotter, U. A. Landscape planning and stress. Urban forestry & urban greening 2, 1–18 (2003).
12. Pretty, J., Peacock, J., Sellens, M. & Griffin, M. The mental and physical health outcomes of green exercise. International journal of environmental health research 15, 319–337 (2005).
13. Storgaard, R. L., Hansen, H. S., Aadahl, M. & Glumer, C. Association between neighbourhood green space and sedentary leisure time in a Danish population. Scandinavian Journal of Public Health 41, 846–852, 10.1177/1403494813499459 (2013).
14. Kaczynski, A. T. & Henderson, K. A. Environmental correlates of physical activity: a review of evidence about parks and recreation. Leisure Sciences 29, 315–354 (2007).
15. Humpel, N., Owen, N. & Leslie, E. Environmental factors associated with adults' participation in physical activity: a review. American journal of preventive medicine 22, 188–199 (2002).
16. Berman, M. G., Jonides, J. & Kaplan, S. The Cognitive Benefits of Interacting With Nature. Psychological Science 19, 1207, 10.1111/j.1467-9280.2008.02225.x ER (2008).
17. Kaplan, S. & Berman, M. G. Directed Attention as a Common Resource for Executive Functioning and Self-Regulation. Perspectives on Psychological Science 5, 43, 10.1177/1745691609356784 (2010).
18. Berto, R. Exposure to restorative environments helps restore attentional capacity. Journal of Environmental Psychology 25, 249, 10.1016/j.jenvp.2005.07.001 (2005).
19. Taylor, A. F. & Kuo, F. E. Children With Attention Deficits Concentrate Better After Walk in the Park. Journal of Attention Disorders 12, 402, 10.1177/1087054708323000 er (2009).
20. Berman, M. G. et al. Interacting with nature improves cognition and affect for individuals with depression. Journal of Affective Disorders 140, 300–305, 10.1016/j.jad.2012.03.012 (2012).
21. Cimprich, B. & Ronis, D. L. An environmental intervention to restore attention in women with newly diagnosed breast cancer. Cancer nursing 26, 284 (2003).
22. Kuo, F. E. & Sullivan, W. C. Environment and crime in the inner city - Does vegetation reduce crime? Environment and Behavior 33, 343 (2001).
23. Kuo, F. E. & Sullivan, W. C. Aggression and violence in the inner city - Effects of environment via mental fatigue. Environment and Behavior 33, 543 (2001).

24. Ulrich, R. S. View through a window may influence recovery from surgery. Science 224, 420–421, 10.1126/science.6143402 (1984).
25. Donovan, G. H. et al. The relationship between trees and human health: Evidence from the spread of the emerald ash borer. American journal of preventive medicine 44, 139–145 (2013).
26. Lasser, K. E., Himmelstein, D. U. & Woolhandler, S. Access to Care, Health Status, and Health Disparities in the United States and Canada: Results of a Cross-National Population-Based Survey. American Journal of Public Health 96, 1300–1307, 10.2105/AJPH.2004.059402 (2006).
27. Dunlop, S., Coyte, P. C. & McIsaac, W. Socio-economic status and the utilisation of physicians' services: results from the Canadian National Population Health Survey. Social science & medicine 51, 123–133 (2000).
28. Mossey, J. M. & Shapiro, E. Self-rated health: a predictor of mortality among the elderly. American journal of public health 72, 800–808 (1982).
29. Idler, E. L. & Benyamini, Y. Self-rated health and mortality: a review of twenty-seven community studies. Journal of health and social behavior 38, 21–37 (1997).
30. Engström, G., Hedblad, B. & Janzon, L. Subjective well-being associated with improved survival in smoking and hypertensive men. European Journal of Cardiovascular Risk 6, 257–261 (1999).
31. Miller, A. B., Baines, C. J., To, T. & Wall, C. Canadian National Breast Screening Study: 1. Breast cancer detection and death rates among women aged 40 to 49 years. CMAJ: Canadian Medical Association Journal 147, 1459 (1992).
32. Vardoulakis, S., Fisher, B. E., Pericleous, K. & Gonzalez-Flesca, N. Modelling air quality in street canyons: a review. Atmospheric environment 37, 155–182 (2003).
33. Amoly, E. et al. Green and blue spaces and behavioral development in Barcelona schoolchildren: the BREATHE project. Environ Health Perspect 122, 1351–1358, 10.1289/ehp.1408215 (2014).
34. City of Toronto, P., Forestry and Recreation. Every Tree Counts: A Portrait of Toronto's Urban Forest. Urban Forestry (2013).
35. Hemery, G., Savill, P. & Pryor, S. Applications of the crown diameter–stem diameter relationship for different species of broadleaved trees. Forest Ecology and Management 215, 285–294 (2005).
36. Troxel, B., Piana, M., Ashton, M. S. & Murphy-Dunning, C. Relationships between bole and crown size for young urban trees in the northeastern USA. Urban Forestry & Urban Greening 12, 144–153 (2013).
37. Rubin, D. B. Multiple imputation for nonresponse in surveys. 81 (John Wiley & Sons, 2004).
38. Rubin, D. B. Inference and missing data. Biometrika 63, 581–592 (1976).
39. Schafer, J. L., Olsen, M. K. & Smardon, R. C. Multiple imputation for multivariate missing-data problems: A data analyst's perspective Multivariate behavioral research 33, 545–571 (1998).
40. Honaker, J., King, G. & Blackwell, M. Amelia II: A program for missing data. Journal of Statistical Software 45, 1–47 (2011).
41. Owen, M., Imai, K., King, G. & Lau, O. Zelig: Everyone's Statistical Software. R package version, 4.1-3 (2013).

42. Rubin, D. B. & Schenker, N. Multiple imputation for interval estimation from simple random samples with ignorable nonresponse. Journal of the American Statistical Association 81, 366–374, 10.2307/2289225 (1986).
43. Hotelling, H. Relations between two sets of variates. Biometrika, 28 321–377 (1936).
44. Johnson, R. A., Wichern, D. W. & Education, P. Applied multivariate statistical analysis. Vol. 4 (Prentice hall Englewood Cliffs, NJ, 1992).

Supplementary information accompanies this paper at http://www.nature.com/srep

PART II

EXPANDING THE URBAN TREE CANOPY

CHAPTER 4

Public Reactions to New Street Tree Planting

RUTH A. RAE, GABRIEL SIMON, AND JESSIE BRADEN

4.1 INTRODUCTION

The arrival of new trees on a city street can transform a space that is both public and private, turning gray sidewalks into green streetscapes. Particularly in densely populated New York City, street trees do not emerge from sidewalks on their own, but their planting requires coordinated efforts and public policies. Through planting a tree at every suitable sidewalk location in this urban environment—on blocks where people live or work—the City of New York is transforming communities, and providing a variety of environmental, social and aesthetic benefits (Figure 1).

Although trees offer benefits to the city overall, the public may not know or understand those benefits. New street trees can elicit positive or negative feelings, and territorial and aesthetic issues can influence

FIGURE 1: These photographs show a New York City street before and after the planting of new street trees, and how the trees can green and soften the streetscape.

perceptions of the value of trees. Trees inserted into the urban environment soften the streetscape and provide aesthetic as well as environmental benefits.

The planting of a tree is significantly different from the arrival of other infrastructure items or static sidewalk furniture such as a light post or street sign. Trees are living things that inhabit the space they are in with a presence—they are iconic woody plants, with archetypal societal implications. People assume they will grow old, become large, reflect seasonal change, and require maintenance and responsibility. Some welcome their arrival with open arms and excitement, while others see their planting as an intrusion into their private space or territory.

This study investigates how some of the public reacted to the planting of these new street trees. Both qualitative and spatial methods were utilized to analyze the opinions communicated to the City of New York which was doing the planting. The examination of emergent correspondence data was rich and grounded in the perspectives of the people. It was not pre-shaped by survey questions but rose up though the open coding of an administrative data set. The public reaction portrayed in the correspondence was both to the new street trees themselves and the planting policies of the City of New York. In order to understand people's reactions, we will begin by describing the new street tree planting process and the public policies that guide the planting process.

4.1.1 NEW STREET TREE PLANTING PROGRAM

New street trees arrive on large trucks, having been pre-dug from fields, and are planted into a sidewalk space that has been cut open by contractors and filled with soil (Figure 2). The new trees are eight to twelve feet high with a trunk girth of approximately three inches. Contractors who plant the trees are supervised by resident engineers during planting, and regulated by contract specifications that contain best practices for healthy street tree planting. The planting locations and tree species have been determined in advance by foresters from the Department of Parks and Recreation's Central Forestry & Horticulture (DPR CF&H) Division to accommodate healthy growth.

FIGURE 2: These photographs show the arrival and planting of new street trees.

Recent large scale municipal planting of street trees in New York City is fueled by the MillionTreesNYC program, proposed as part of Mayor Bloomberg's PlaNYC in 2007. PlaNYC's goal is to create a greener, greater NYC, with 127 initiatives intended to improve the physical city; impacting land, water, transportation, energy, air and climate change (City of New York 2007). The plan's focus is to provide for sustainable improvements to NYC, which requires new levels of collaborations and substantial resources. The DPR CF&H Division, in collaboration with non-profit and other partners, will plant one million trees by 2017. These plantings on public and private property have the potential to increase the overall tree canopy cover for New York City, which was estimated at 24% in 2001 (Grove et al. 2006).

The development of an urban forest requires significant public investment, and MillionTreesNYC combines both public and private funding sources. The Parks Department will plant sixty percent of those one million trees in public space (220,000 on streets with an additional 380,000 trees in woodland areas or open park space), while forty percent will be planted by the City's partners (New York Restoration Project and other organizations) on public and private land (www.milliontreesnyc.org; Stephens 2008).

The Mayor has pledged to fill all available sidewalk spaces with street trees by 2017 to raise the street tree stocking level from 74% to 100% (City of New York 2007). Since the area between the curb and the property line belongs to the city, the plan is to create a ribbon of green along this gray public space. New street trees will green the cityscape, and beautify the public realm to improve the experience of every pedestrian. Between 2007 and 2009, the DPR CF&H Division has already planted 53,235 new street trees.

Historically, the DPR CF&H Division planted trees on an individual request basis. This meant that citizens could request a free tree planting in front of their property, which was fulfilled on a first-come first-served basis, since the demand could often exceed the supply of trees. In addition to individual requests, foresters would also identify additional locations for street tree plantings in front of properties where no tree request had been made, and building owners were given the option to refuse the tree planting. Under this method, one or two trees might be placed on a block at one time, and trees could also be planted based on an unequal distribution of requests.

With the beginning of PlaNYC and MillionTreesNYC, there was a major policy shift in how street tree planting was done. PlaNYC funded the capital budget to provide for the large-scale volume planting of new street trees. This led to the creation of the block planting program and the development of a methodology to assess and target those neighborhoods in the greatest need of new street trees. The sections of the city with low street tree stocking level and high population density receive prioritized planting under the program, ensuring that tree benefits are maximized and the scope of the initiative reaches all citizens by the scheduled conclusion of MillionTreesNYC in 2017.

Along with this new program, the City enforced its legal authority over the sidewalk and implemented a planting policy that no longer allowed building owners the ability to deny a suitable tree planting in the public right-of-way. Trees are still planted to fulfill requests from citizens, and approximately thirty to forty percent of trees planted are in response to individual requests citywide. However, the majority of new street trees planted by DPR CF&H follow the policy priority of mass block planting. Block planting brings trees and their benefits to neighborhoods that previously had few or no trees, while also making significant strides towards accomplishing planting goals.

4.1.2 STREET TREE BENEFITS AND CONCERNS

The accrued benefits of street trees have been quantified and translated into financial value (Peper et al. 2007; Nowak et al. 2007; McPherson et al. 2007). As of 2005, the City had 592,130 street trees that were estimated to provide approximately $121.9 million in annual gross benefits (Peper et al. 2007). Planting along streets and in parking lots provides additional benefits beyond those that come from planting in parks due to the shade of structures (Peper et al. 2007).

The detailed analysis of the New York City urban forest by the U.S. Forest Service was used by Parks Department's Commissioner Adrian Benepe to secure $400 million for tree planting from the city budget

(McIntyre 2008). In this calculation, both the environmental and aesthetic benefits that the urban forest produces for the community are linked to the quality and extent of New York City's canopy cover. Fifty-seven percent of the benefits are environmental and include the capture of storm water runoff, energy savings, air quality improvement and the reduction of carbon dioxide (Peper et al. 2007; Nowak et al. 2007; McPherson et al. 2007). The other forty-three percent of the benefits relate to beautification, the associated aesthetic values and annual increases in property value (Peper et al. 2007; Nowak et al. 2007; McPherson et al. 2007).

Several studies have assessed the social benefits of urban and community forestry programs (Westphal 2003; Kuo 2003). Research found that outdoor spaces with trees were used more frequently than spaces without trees, and that this facilitated interactions among residents that fostered more sociable neighborhood environments and stronger neighborhood social ties (Kou et al. 1998). Views of trees provide restorative experiences that ease mental fatigue (Kaplan and Kaplan 1989). By making residential outdoor spaces more vital, trees can contribute to the functioning of a healthy community (Kou 2003; Kou et al. 1998).

Urban forestry programs often involve community-based greening activities (Wolf 2003). People who either planted their own tree or participated in a tree planting program reported greater satisfaction and were more likely to think the tree improved the yard and the neighborhood (Summit and Sommer 1998). If volunteers plant trees themselves their relationship, attitude and satisfaction with the tree planting is substantively different than those planted by a municipality using hired contractors (Sommer et al. 1994).

Trees have many meanings for people. The connection between human beings and trees is strong, for trees can shape both individual and collective identities (Sommer 2003). Human beings derive pleasure from trees (Lewis 1996) and trees can also represent personal, symbolic, and religious values (Dwyer et al. 1991). They can commemorate people who have passed (Svendsen and Campbell 2005; Tidball et al. 2010) or children just born, for they have spiritual value and longevity. Trees are more than just a decorative feature in the landscape—they have the ability to transform it over time at both a physical and psychological level.

The aesthetic aspects of trees have also been found to be important. Several studies have found that there are visual preferences for a certain size, shape or form of a tree (Williams 2002; Schroeder et al. 2006). The majority of reported positive features of street trees were found to be related to aesthetic considerations such as being pleasing to the eye, the giving of shade, enhancing the look of a garden or home, and making the neighborhood more live-able (Gorman 2004). These intangible benefits of aesthetics had the strongest correlation with the overall assessment of a street tree right outside the home (Schroeder et al. 2006). Issues of comfort (shade) and appearance play more of a role in the decision to plant trees than do concerns about environmental benefits or energy savings (Summit and McPherson 1998). Trees, by adding softer natural elements to a city, also enhance the public's impression of the visual quality of cities (Wolf 2008). Beautification is one of the most frequently cited reasons for why people plant trees (McPherson 2007)

However, trees do require maintenance and imply responsibility. They drop leaves and can damage sidewalks. Studies have found that urban trees can cause annoyances and involve liability issues. Trees can be considered to be messy or dirty by some (Sommer 2003). Gorman (2004) found that complaints about trees had to do with power line interference, sidewalk damage, and visibility blockage. There are issues with actual root damage to property, falling leaves or limbs, general debris, or the reduction of personal safety by limiting visibility views from the property (Schroeder et al. 2006). The planting and management of trees can conflict with other elements of the urban infrastructure such as sewers and sidewalks (McPherson et al. 2007).

4.1.3 THE SIDEWALK GREY ZONE

In New York City all trees growing in the public right-of-way, along streets and in parks, are under the jurisdiction of the Parks Department, which manages about half of the City's 5.2 million trees (Nowak et al. 2007). The City of New York owns the space between the curb and the building owner's property line, but the owner is responsible for the maintenance of the sidewalk. New York City law requires property owners to repair the sidewalk adjacent to their properties at their own cost[1]. The Department of Transportation can issue violations for sidewalk defects for public safety reasons (New York City Department of Transportation, 2008). The legal responsibilities for liability related to sidewalks, tree roots and tree wells has changed over time and by residential property type, so that liability and ownership can be blurred (Kaye et al. 2009). The collective history of New York's tree and sidewalk laws reflect competing interests and conflicts between property owners and city agencies.

The greening of cities through the installation of trees into sidewalks is not inherently controversial, yet it can create conflict because of people's territorial instincts, and vagueness in legal issues defining the responsibilities of the city and citizens. Sidewalks are seen as public spaces that should encompass diversity and have multiple functions, yet these places can also be contested terrains (Loukaitou-Sideris and Ehrenfeucht 2009). Even though street trees are generally desirable, they elicit varied responses from urbanites who want different things from public space.

The planting of trees on residents' streets and in front of homes raises issues of territoriality and place attachment. Human territoriality involves the drive to establish control over physical spaces and involves the demarcation and defense of space against territorial invasion (Brown 1987; Taylor 1988; Sommer 2004). Human territoriality is linked to concepts of personalization and privacy (Sommer 2004). Territorial emotions can involve a positive emotional bond to a place and belief that they should have control over the condition of the site and who should be there, or a negative emotional reaction to changes in conditions or users of an area (Wickham and Zinn 2001). Territorial behavior is strongest when considering individuals or small groups and when the spatial focus is on specific small scale locations. Territorial functioning refers to sentiments, cognitions, and behaviors that are highly space specific and represent transactions concerned with the management, maintenance, legibility, and expressiveness of person-place transactions (Taylor 1988).

Types of territories exist along dimensions of occupancy and psychological centrality (Brown, 1987). Primary territories are locations central to people's lives and typically are homes. Outdoor residential settings, including front yards, sidewalks, driveways, backyards, and the street itself can also have strong centrality (Taylor 1988). The planting of trees on the streets where people live and adjacent to homes can affect the 'lifespace' of an individual, since in going to and from home people must transverse these places. Residences are inextricably linked with the area right outside the door, not only physically but psychologically as well.

Primary territories allow for order, predictability and control, as well as the expression of a sense of identity (Brown, 1987). People often 'mark' or personalize the areas around their homes leaving behavioral traces such as decorations or signs of upkeep. Territorial behaviors also include boundary control efforts to manage the access and activities of others. Territorial cognitions include the perceptions of and affect toward a place including issues of responsibility, caring and the association or appropriation of a place (Taylor 1988).

Human territorial emotion is closely related to place attachment at the affective level (Wickham and Zinn 2001). Place attachment involves human bonding to a place, which has affective, cognitive and behavioral components (Low and Altman, 1992; Manzo 2005; Proshansky et al. 1983). A physical space becomes a place when it encompasses memory, attachment, and identity. Places have a geographic location and material form, but they are also invested with meaning and value by ordinary people (Gieryn 2000). Territoriality is intimately related to how people use land, how they organize space, and how they give meaning to a place (Sack 1986). A sidewalk where a new street tree is planted may not be just a physical space but can also be a place that has meaning to people. Since place attachments are holistic but can operate in the background of awareness, they become more fully recognized when they have been disrupted (Brown and Perkins 1992).

The sidewalk belongs to the City and is a public right of way, but not every resident wants a tree planted there regardless of the public benefit. Sidewalks are both public and private spaces. They must allow for public access, but can also evoke feelings of personal ownership and territoriality. Although trees physically transform the grey infrastructure of sidewalk into a green space, the sidewalk is a literal, figurative, and psychological grey zone.

4.2 RESEARCH METHODS

This study examines the results of the content and spatial analysis of the correspondence data and its relationship to the operational policies and procedures that guide New York City's municipal street tree planting program. Qualitative analysis of the administrative correspondence data set examined the public perceptions and concerns related to the MillionTreesNYC program. Spatial analysis explored the relationship between the block planting locations of new street trees and the locations of the citizen correspondence regarding both requests for new street trees and complaints.

In late 2006, DPR's CF&H Division created a database system to track and log correspondence from receipt to resolution. This database made this study possible with its capacity to record and conduct analyses of detailed qualitative and quantitative data. The original intention of the qualitative coding was to easily identify similar themes in the correspondence from the public as it became apparent that the same topics were being repeatedly addressed. Categorization of concerns increased the efficiency of the responses to the public, helped in the creation of standardized template responses, and improved reporting.

The qualitative analysis of the administrative correspondence data set involved the open coding of text from 311 call transcriptions, letters, and emails received by DPR CF&H Division between 2007 and 2009. The categories identified were not solicited or manipulated by any sort of directed questioning, but were instead determined from the open-ended coding and content analysis of this existing correspondence data set. There are limits to this data set since it was not research guided by a survey tailored to testing a certain hypothesis; demographic information was not collected, nor was this from a random sample of the population. Instead the correspondence analyzed reflected the concerns of individuals who were self-selected in that they chose to contact the City of New York concerning new street tree plantings.

The open coding of this correspondence allowed the perspectives of the people and grounded theory to emerge. As the core categories were identified and dimensionalized through open coding, more axial and selective coding began (Strauss and Corbin 1990). In order to code this correspondence by category, we created a multifaceted coding system that included a variety of primary categories and an array of more detailed subcategories. These subcategories gave dimensions to the primary categories and made them more robust. Through further comparisons and examination of relationships, the categories used were further refined and collapsed (eighteen primary categories were collapsed to sixteen, and several subcategories were also combined). Selective coding and frequency analysis, combined with operational and policy analyses, identified various paradigms and patterns that explained the phenomenon of public reactions to new street tree planting.

All items of correspondence received were coded using a multi-level system[2]. A type classification was assigned to each correspondence item received followed by the identification of primary categories and related subcategories to specify the precise subject matter of the inquiry. In the majority of correspondence items, 87%, only one issue of concern was noted, but 10% had two categories and 3% contained three or more issues. Sixteen primary categories and numerous associated subcategories were identified from the qualitative content analysis of correspondence received by CF&H from 2007 through 2009. Frequency analysis of these primary categories found that 81% were comprised of seven New Street Tree (NST) categories.[3] For the purposes of this study, further analysis was done only on these seven New Street Tree primary categories of which two were Service Requests while five were Complaints. The seven NST primary categories totaled 3,838 of which the NST Complaints subtotal was 2,561 and the NST Service Requests subtotal was 1,277 (Table 1).

Spatial analysis of the data using geographic information systems (GIS) was also performed. The development of geospatial tools has contributed to urban forest management by enabling rapid analysis of current data (Ward and Johnson 2007). Analyzing the distribution of planting requests, citizen complaints, and block planting progress allowed for the comparison of content analysis categories, new street tree requests, and the locations of new street tree plantings by the Parks Department. In particular, block planting locations could be spatially compared to the public reactions to new street tree planting. The GIS method brought together

Table 1. Communalities for the variables based on the canonical correlation analysis.

Correspondence Type and NST Primary Categories	Total	Percent of Total	Percent of Type
NST Complaint			
Placement Objection	859	22%	33%
Policy Objection	606	16%	24%
Maintenance Objection	439	11%	17%
Resultant Damage	358	9%	14%
Process Objection	299	8%	12%
Subtotal	2,561	66%	100%
NST Service Request			
New Tree Request	636	17%	50%
New Tree Conditions	641	17%	50%
Subtotal	1,277	34%	100%
TOTAL	3,838	100%	

both the operational prioritization policy and the actual tree planting sites with the content analysis research findings.

As part of DPR CF&H Division's ongoing GIS program, planting locations are tracked and updated every season at both a street block segment and individual tree location level. Individual trees are tracked in parts of a Forestry Management System database that includes a spatial component. Block planting street segment locations are provided by the foresters and input directly into the GIS as line segments. The existing data on block planting locations was analyzed against two additional data sets created from the correspondence data. The first was a density of tree requests raster layer. Between 2007 and 2009, 14,908 requests for new street trees were received by 311 and transmitted directly into a Forestry Management System database utilized by DPR CF&H. ArcGIS's Spatial Analyst extension was used to transform these addressed-based point requests into a raster density layer. The second analysis, using the same method, generated a density of 2,561 complaints raster layer for the same time-period using new street tree complaints from the DPR CF&H Division correspondence database.

4.3 FINDINGS AND DISCUSSION

4.3.1 SPATIAL ANALYSIS OF NEW STREET TREE PLANTING

Figure 3 shows the result of the 311 New Tree Request data raster analysis. This map depicts where citizens have requested new street trees and where block planting has taken place. The block planting segments include plantings since the inception of MillionTreesNYC in 2007 through 2009. Block planting segments are tracked every planting season using data provided by foresters. After each planting season, foresters submit field maps with block planted segments highlighted. This information is incorporated into the GIS layer and is shown as purple lines on the map. The green areas show the spatial density of the 14,908 new tree requests from 311 during the same time period. Darker green areas indicate more new street tree requests received directly by the New York City 311 Customer Service Center. The map shows that under the block planting program, the Parks Department's CF&H Division is block planting in areas where citizens have generally not requested trees from 2007 through 2009.

Figure 4 also shows a map that depicts where block planting has taken place between Fall 2007 through Fall 2009. This time, however, block planting is plotted against the density of 2,561 citizen complaints CF&H received from 2007 through 2009. The density of these complaints are shown in blue, with darker areas indicating more complaints. As can be seen, the highest density of complaints cluster around the block planting locations shown in purple and display the high volume of citizen complaints that are coming from areas of recent block plantings. Some light blue areas that are not in proximity to block planting areas are reactions to individual tree plantings. These show as higher density due to multiple complaints made by one person or multiple complaints made by several people who live relatively close together. The specific content of these complaints are discussed in the New Street Tree Complaints section.

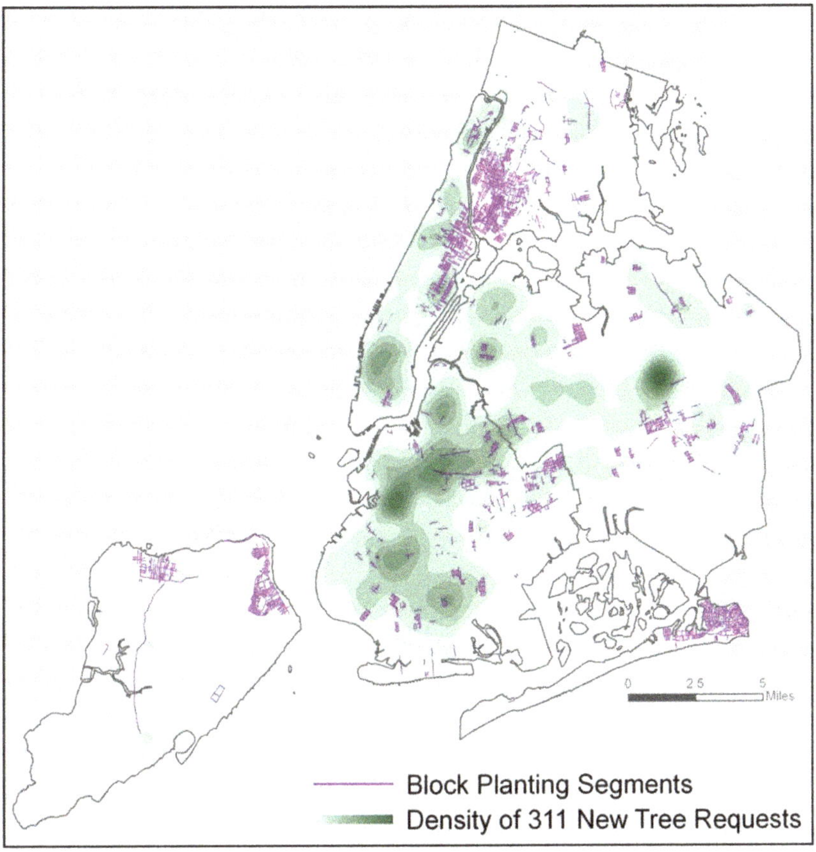

FIGURE 3: Map of 311 New Tree Requests Density and DPR CF&H Block Planting Locations for New York City from 2007 through 2009.

4.3.2 CITIZEN CORRESPONDENCE OVERVIEW AND VOLUME

This study investigated the content of letters, emails, and transcriptions of calls from 311 received by the Parks Department's Central Forestry and Horticulture Division between 2007 and 2009. At the broadest level, each item of correspondence is assigned to one of five basic type groups:

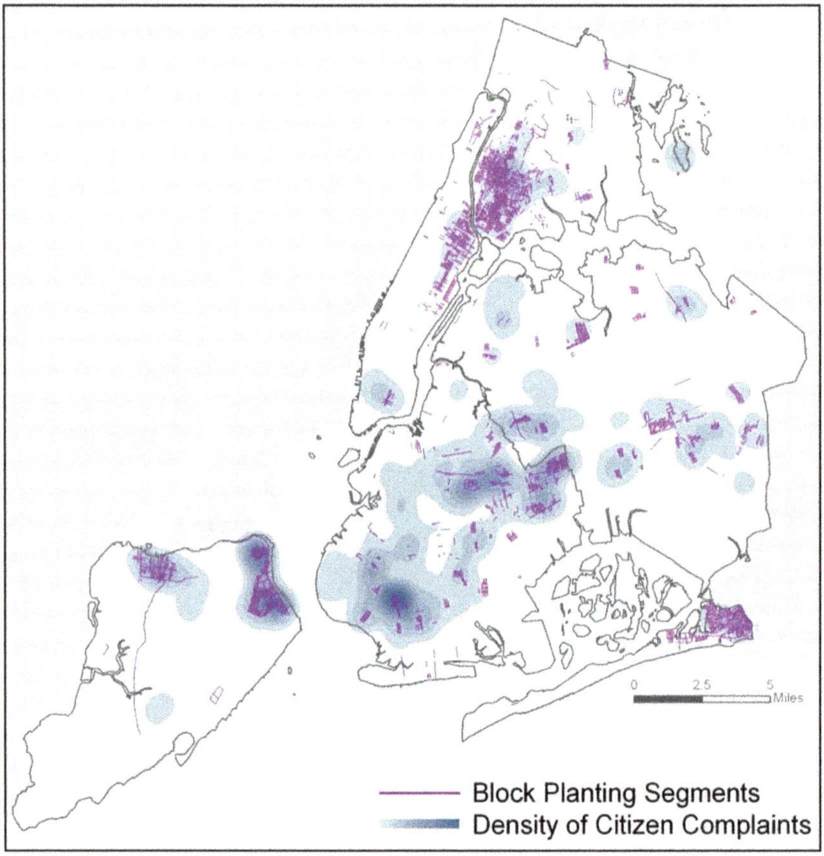

FIGURE 4: Map of DPR CF&H Citizen Complaints Density and Block Planting Locations for New York City from 2007 through 2009.

Complaint, Service Request, Information Request, Recommendation, or Thank You. Table 2 below shows the frequency of the total items of correspondence by type from 2007 through 2009. Complaints are the most frequent type of correspondence received at 57%, followed by Service Requests at 29%.

Table 1 also shows the three-fold increase in the total items of correspondence received by the CF&H Division from 2007 to 2008 and 2009. In reaction to MillionTreesNYC's street tree planting initiative, CF&H

has witnessed a vast increase in the amount of citizen correspondence since the MillionTreesNYC program's inception in 2007. The increase can be attributed to the expansion of the street tree planting program, improved public accessibility and awareness of New York City's 311 Customer Service Center services, and the efficiency of the 311Center's linkage to city agencies.

Figure 5 shows both the total items of correspondence received from 2004 until 2009 and the increase in the amount of street trees which were being planted in the same year. The amount of correspondence increased dramatically with the increase in street tree planting and the inception of the Million-TreesNYC program in 2007, and is most noticeable in 2008 and 2009.

4.3.3 CONTENT ANALYSIS OF NEW STREET TREE CORRESPONDENCE

Table 2 separates the seven New Street Tree (NST) primary categories by type of correspondence and gives the totals and relative percentages for each primary category as percent of the overall total and within each category type. Of those seven primary NST categories, five were Complaints (66%) and two were Service Requests (34%). Under the Complaint type these include the primary categories (in order of frequency) of Placement Objection, Policy Objection, Maintenance Objection, Resultant Damage and Process Objection. Service Request types include the primary categories of New Tree Requests and New Tree Conditions. Following will be a discussion of each of these primary categories and the subcategories they contain separated by type.

4.3.4 NEW STREET TREE COMPLAINTS

New Street Tree Complaints are objections to elements of the tree planting process, from general dissatisfaction with the mandated new program to specific rationalized objections to a planting at a given location, or stages of that planting process. Concerns over the placement of a particular new tree planting based on surrounding site conditions, objection to the tree

Table 2. Type of Correspondence by Year.

Correspondence Type	2007	2008	2009	Total
Thank You	24	15	28	67
Recommendation	51	19	32	102
Information Request	94	137	153	384
Service Request	190	241	764	1,195
Complaint	225	1,211	924	2,360
TOTAL	584	1,623	1,901	4,108

based on perceptions of future maintenance responsibilities, dissatisfaction with the agency's notification measures, or the quality of the work performed by the landscape contractors in planting, are all examples of common correspondence defined as Complaints. New Street Tree (NST) Complaints were coded into five primary categories: Placement Objection, Policy Objection, Maintenance Objection, Resultant Damage, and Process Objection. Table 2 depicts the total for each of these NST primary categories, and their percentage within the total of the New Street Tree Complaint type. The tables below (Tables 3, 4, 5, 6 and 7) show the subcategories that make up these primary complaint categories.

Placement Objection at 33% was the largest primary category of complaint (see Table 3). These subcategories of placement objections are considered to be logical appeals against the particular placement of a given planting location or situation. The subcategories of concern are often not in complete opposition to trees or their presence in the urban environment, but believe that a given location for a new tree is unpractical or unsuitable for tree planting because of existing infrastructure or site usage. Within Placement Objection, utility line concerns are most prevalent (33%). This variety of objection is typically brought by residents that fear a utility line, whether it is gas/electric/water/sewer, will suffer damage because of the tree planting process[4] or the tree's growth at a given location. The next most popular subcategory was complaints against the street tree planting

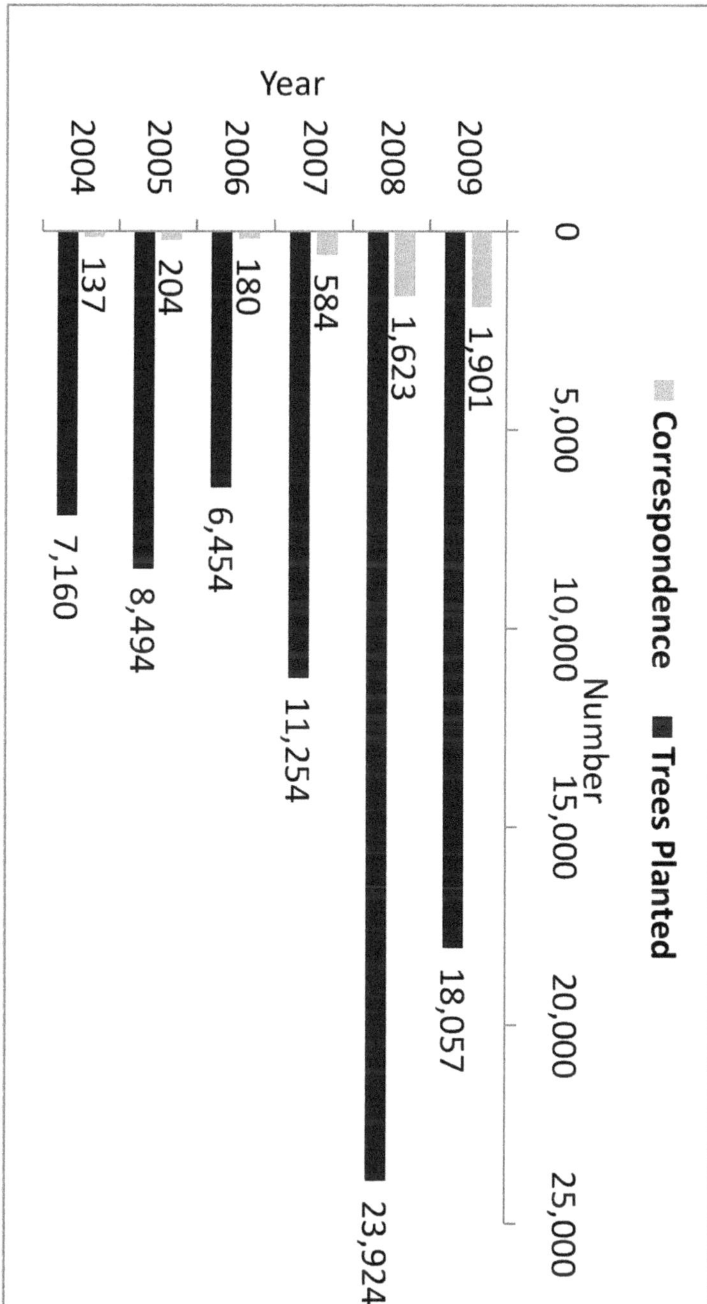

FIGURE 5: Total Items of Correspondence and Total Number of New Street Trees Planted from 2004 through 2009.

Table 3. New Street Tree Complaint—Placement Objection Subcategories.

Placement Objection Subcategories	Total	Percent
Utility Line Disturbance	288	33%
Sidewalk	118	14%
Driveway	76	9%
Existing Tree	70	8%
Business Disturbance	51	6%
Private Property	48	6%
Visibility Interference	45	5%
Disability Concerns	44	5%
Infrastructure Conflict	41	5%
Miscellaneous	40	5%
Shade	38	4%
TOTAL	859	100%

because of the damage to the sidewalk or a narrowed sidewalk (14%). This may be related to a resident's sense of ownership over such a publicly used space, particularly in an urban setting where walking and public transit are the most common forms of transportation. A variety of complaints relate to the perception that the new street trees are too close to existing infrastructure items or private property. Objections against placement can also be based on disturbance to usage patterns, visibility or special circumstances. Proper placement may be the most difficult obstacle to planting in a highly dense urban setting—it relates to the ambiguous public/private nature of the sidewalk space, issues of territoriality and misunderstanding of the procedural processes and guidelines followed during planting.

Policy Objection consists of approximately 24% of the total NST Complaints received (Table 4). The majority of these are general refusals of new trees in front of a given property. At 57%, these general refusals originate from citizens rejecting a planned planting without supporting reasons or explanations for their complaint—they simply state they do not want a tree. This subcategory assumes a general dissatisfaction with the planting policies of the Parks Department and their public right of way authority to plant at given sites without the expressed permission of adjacent property owners. It also depicts how the planting of street trees can evoke issues of territoriality and control. A total of 28% of complaints are objections to planting based on lack of notification prior to planting; 18% complain of a general lack of notification, while 10% of these complaints are objections to the cut in their sidewalk or were reported by property owners who had recently paved their sidewalks. Complaints of poor notification indicate either actual property ownership or a sense of ownership over this shared sidewalk space.

Maintenance Objections comprised 17% of total NST Complaints (see Table 5). While not the largest subcategory of complaint, Maintenance Objections are linked to sentiments of ownership and responsibility for the sidewalk. Whether because of the proximity to their front door, feelings of territoriality or their concerns regarding liability, this is typically one of residents' most adamant objections. Most Maintenance Objection complaints express an apprehension about future responsibilities for tree care. These include the raking of leaves and watering (36%), followed by the fear of future sidewalk or foundation damage caused by a growing tree and its roots (29%). The motivations behind an objection can be linked directly to the laws of the municipality and the public's interpretation of the statutes. Many citizens also express fear that the trees may become receptacles for dog waste and litter (24% in total), creating an unpleasant experience for the resident and perceived added responsibilities to keep the area clean because of the risk of a Department of Sanitation violation and fine.

Resultant Damage represents 14% of total NST complaints (see Table 6). These are issued by residents who are dissatisfied with the quality of the tree planting based on damage that occurred to the surrounding location. The majority of these are complaints of damage to the curb or sidewalk adjacent to the tree planting (55%). Also, 26% are complaints against the planting contractor for debris or material left on site, including packing materials or excess concrete from sidewalk excavation.

Process Objections accounts for 12% of NST Complaints (see Table 7). These citizens take issue with the logistical stages of planting operations

Table 4. New Street Tree Complaint—Policy Objection Subcategories.

Policy Objection Subcategories	Total	Percent
General Refusal	343	57%
Notification-General	111	18%
Notification-New Sidewalk Cut	57	10%
Miscellaneous	52	8%
Pit Size	43	7%
TOTAL	606	100%

Table 5. New Street Tree Complaint—Maintenance Objection Subcategories.

Maintenance Objection Subcategories	Total	Percent
Tree Care	157	36%
Future Sidewalk & Foundation Damage	128	29%
Dog Waste/Litter	106	24%
Prior Experience with Property Damage	48	11%
TOTAL	439	100%

Table 6. New Street Tree Complaint—Resultant Damage Subcategories.

Resultant Damage Subcategories	Total	Percent
Sidewalk/Curb	197	55%
Debris Left at Site	93	26%
Utilities	38	11%
Private Property	30	8%
TOTAL	358	100%

Table 7. New Street Tree Complaint—Process Objection Subcategories.

Process Objection Sub-Categories	Total	Percent
Unplanted Excavated Tree Pit	163	55%
Species Assignment	57	19%
Contractor Misconduct	34	11%
General	30	10%
Confiscation of Sidewalk Decorations	15	5%
TOTAL	299	100%

from sidewalk survey and markings, to excavation, to planting. The majority of these complaints may stem from the perception of liability or the recognition of an obvious hazard caused by the planting operation. The majority of the concerns (55%) were because of excavated sidewalk plots left unplanted. Pre-excavation is a common stage of the NYC street tree planting process as the planting contractors often pre-excavate planting sites to expedite the installation of trees. The practice of pre-excavation requires the contractors to secure the opened site with enough soil to bring the area to grade with the sidewalk. In some cases, soil settles below the sidewalk grade, or the citizen may be uninformed of the sequences of the planting process. In other cases, residents take issue with the type of tree species chosen by the forester, often asking for a different variety to be selected (19%). These residents are accepting of the possibility of tree planting at this site, but would like more control over the planting since they expect the tree to become a part of their daily lives. The confiscation of sidewalk decorations (5%) complaint relates to the personalization and marking of territory that can be disrupted by the process of planting a street tree.

All NST Primary Complaint categories included a subcategory that addressed public reaction to or concern over the sidewalk in relation to the trees being planted. Complaints about new street tree planting are often motivated by the public's perception that the sidewalk is owned by its citizens, particularly those citizens that live, work, or own adjacent property. These subcategories reflect the sidewalk as a place that is both a public and private space. Ownership conflicts and responsibility concerns are evidenced, as are misunderstandings of the planting process and issues of territoriality.

4.3.5 NEW STREET TREE SERVICE REQUESTS

NST Service Requests are correspondence where the public requests specific actions regarding new street tree planting. This can include requests for new tree plantings, or maintenance on a recently planted tree. These requests for service are typically public reports of tree conditions or tree pits that require investigation, inspection and action. In the case of new street tree planting, these are largely positive categories that depict a desire and concern for trees, and can gauge civic support for the citywide greening program. NST Service Requests include people following-up on the status of their tree requests, making new tree requests under special circumstances, or asking for additional work to be performed in the maintenance of a recently planted tree. The two major subdivisions of correspondence of the Service Request Type, at 50% each, were New Street Tree Requests (636) and New Tree Conditions (641) (see totals and subcategories in Tables 8 and 9).

Separate from the DPR CF&H correspondence database, there are 311 New Tree Requests which are transferred directly from 311 into the DPR CF&H Division's Forestry Management System database for assessment by foresters. Between 2007 and 2009, 14,908 311 New Tree Requests were received in this manner, and formed the basis for the spatial analysis for the 311 New Tree Request map discussed earlier (Figure 3). In 2007 there were 639 received via 311, but in 2008 and 2009 there was a dramatic increase, and over 7,000 requests were received each year. This increase was due to the visibility of the MillionTreesNYC campaign and the accessibility of the 311 Customer Service Center.

Some New Street Tree Requests (646) are transmitted to the Parks Department CF&H correspondence liaison by the 311 intake operator because

Table 8. New Street Tree Service Request—New Tree Requests Subcategories.

New Street Tree Requests Subcategories	Total	Percent
Individual Planting-Initial/ Status	342	53%
Block Planting-Initial/ Status	114	18%
Damaged Tree Replacement	109	17%
Commemorative Tree Planting	38	6%
Tree Request Rejection/ Cancellation	33	6%
TOTAL	636	100%

Table 9. New Street Tree Service Request—New Tree Conditions Primary Category, Tree Health and Tree Pit Subcategories.

New Street Tree Conditions	Total	Percent
Tree Health Subcategories		
General (Unhealthy)	229	36%
Vandalism	55	9%
Tree Stakes	50	8%
Watering Needed/ Gator Bags	36	5%
Miscellaneous	27	4%
Tree Health Subtotal	400	62%
Tree Pit Subcategories		
Paving Stones	103	16%
Tree Guards	80	13%
Hazardous Pit/Maintenance	58	9%
Tree Pit Subtotal	241	38%
TOTAL	641	100%

of their unique nature, or were transmitted directly to DPR's CF&H by letter or email, and these are reflected below in Table 8. The majority, at 53%, relate to the individual planting requests or status inquiries, while another 18% have to do with initial or status block planting requests (since sometimes people request that their entire block be planted). These can also be new service requests with special features that need attention such as when a requested new street tree arrives with damage and requires replacement (17%) or the planting of a commemorative tree (6%).

New Street Tree Conditions are comprised of two separate subcategories with multiple concerns listed within each (Table 9). The majority of requests for service related to concerns about the health of a newly planted tree (Tree Health at 62%), while 38% relate to issues surrounding the Tree Pit (which is the earthen area surrounding the street tree). The major Tree Health subcategories (36%) are reports of new trees which generally look unhealthy and need help. There are also reports of incidences of vandalism (9%) against trees, or about missing tree stakes (8%), which support the trees when they are growing. The NST Condition category also includes requests for work on the Tree Pits, including the installation of paving stones around the perimeter (16%), the installation of tree guards (12%), or the correction of other perceived tree pit hazards (9%). All of these Service Requests show public interest in a recently planted area and imply concern and responsibility for the newly planted street tree.

4.4 DISCUSSION AND FUTURE RESEARCH

The correspondence that came to DPR's Central Forestry and Horticulture Division from 2007 to 2009 reflects the public's concerns and response to new street trees being planted. Particularly in 2008 and 2009, the volume of correspondence about new street trees grew dramatically in response to the increase in street tree planting initiated in 2007 by PlaNYC and the MillionTreesNYC program. Two-thirds of the correspondence categories about these new street trees involved complaints or objections to new street trees, while one-third were service requests either related to requests for a new street tree or concern for trees that had just been planted.

The MillionTreesNYC campaign has an extensive public outreach component, advertising the tree planting program and its goals and benefits on subways, bus stops and in the media. Yet the public may still not know about, understand, or appreciate the benefits of new trees. Even though trees are substantive living things that have meaning for people and can foster feelings of attachment, they can also involve responsibility, care and maintenance. Maintenance objections were the third most prevalent category of complaint about new street trees.

The demand for new street trees and the popularity of the street tree planting program is portrayed in the spatial analysis of the almost 15,000 311 New Tree Requests. Even though new street trees are still being planted in response to individual requests, the MillionTreesNYC priority is to plant street trees by block in order to target the areas of the city with the most people and least trees. This spatial analysis showed that the street tree block planting areas were not necessarily being planted where people had made requests.

GIS analysis also showed that the highest density of citizen complaints were coming from areas of recent block plantings. Planting individual trees increases green infrastructure throughout the landscape, but block planting in particular transforms grey sidewalks of entire streets into ribbons of green. Yet block planting, and sometimes even individual tree planting, can sometimes happen without residents being aware the trees are coming. Some welcome this planting, while others are wary.

Objections to placement location was the biggest complaint about new street tree planting, followed by policy objections where people did not want a tree or had not been notified in advance before their sidewalk was cut or the tree was planted. Urban residents can be bothered by the placement of trees in sidewalks, a literal grey zone that is both a public and private space, especially if they did not ask for them. Despite the fact that the public benefits should outweigh these personal disturbances, people have a sense of territory about their homes and streets. Even though the sidewalk is legally a public right of way with government jurisdiction, residents can have a psychological sense of ownership over this place that can have personal meaning.

Involvement in the planting process could help to transfer a citizen's sense of ownership over the sidewalk through giving them more investment in new street trees. However, given the scale and complexity of the Parks Department's citywide planting of new street trees, a large scale citizen involvement with the planting of street trees would be difficult to manage. MillionTreesNYC does have biannual volunteer planting days, but these involve the planting of trees in parks citywide (City of New York 2010). They also have a website5 that provides educational publications including instructions on tree care and an explanation of all the steps in the street tree planting process.

The DPR CF&H Division also conducts public outreach about its upcoming block street tree planting activities via the posting of block planting posters and flyers. Additional targeted education on tree benefits and expanded notification of planting processes and procedures, particularly in advance in targeted block planting areas, could increase public acceptance of the new street tree planting. If residents were more aware of what was about to happen to their street and sidewalks, they might be more receptive to the new street trees.

The block planting of street trees actually offers an excellent opportunity for a natural experiment. Control groups could be designated for communities targeted for block planting: certain blocks would receive more intensive education on tree benefits and the street tree planting process, while other adjacent blocks would receive no or less notification and education. This method would investigate the impact of education and outreach on reactions to new street tree plantings by the City.

There are many opportunities to conduct additional research that further investigates the public's reactions and perceptions of new street trees. A random sample could be stratified by demographic and socio-economic variables in order to investigate differences in the perception of new street trees by neighborhood. Photographic methods could be utilized to assess residents' perceptions of changes in the aesthetics of the streetscape before and after planting. Using spatial analysis to examine differences in public perception by boroughs, neighborhoods, housing types and home ownership would also be valuable.

Given the dual public and private issues surrounding the sidewalk, it is likely that property ownership would be a significant factor. Public policies concerning liability for the sidewalk have changed over time, but regardless of the actual law, the public understanding of this complicated city policy will continue to be murky. A future study could investigate how sidewalk maintenance liability laws impact the public perceptions and reactions to street tree plantings.

Focus groups held in affected neighborhoods could identify areas of concern not revealed in the analysis of this self-selected sample of people who corresponded with the City about their newly planted street trees, and could also identify the more positive reactions to new street tree planting. This information, along with the categories of concern uncovered through this study, could lead to the development of a robust survey instrument that could be administered to targeted areas to evaluate the full spectrum of responses to new street tree planting. This research would lead to a fuller and more comprehensive understanding of both the positive and negative aspects of people's reactions to new street tree planting.

FOOTNOTES

1. There are some exceptions to this, especially in relation to one, two or three family residential properties.
2. The qualitative coding majority of was performed by the correspondence liaison. A trained intern assisted with coding after the categories were well defined, and a comparison of their coding with the liaison found a 98% agreement of selected categories.
3. Nine of the general primary categories consisted of only nineteen percent of the total items of correspondence received: these consisted of Appreciation, Donations/Solicitations, Greenstreets, Insects, Mature Tree Maintenance, Permits, Public Health, Research and Miscellaneous.
4. Planting guidelines require that before work begins the utility companies mark the locations of underground lines on the sidewalk to ensure that contractors are aware of their presence while excavating the planting sites.
5. http://www.milliontreesNYC.org

REFERENCES

1. Brown, B. 1987. Territoriality, pp. 505-532. In Stokols, D. and I. Altman. (Eds.). Handbook of Environmental Psychology. John Wiley & Sons, New York.
2. Brown, B. B. and D.D. Perkins. 1992. Disruptions in place attachment, pp. 279-304. In Low, S.M. and I. Altman. (Eds.). Place Attachment. Plenum Press, New York.

3. City of New York. 2007. PlaNYC: A Greener, Greater, New York, Mayor Michael R. Bloomberg, The City of New York. http://www.nyc.gov/html/planyc2030/html/home/home.shtml (accessed 05/10/2010).
4. City of New York, 2010. PlaNYC Progress Report 2010, a Greener, Greater New York. 2010. Mayor's Office of Long-Term Planning & Sustainability, City Hall, New York, NY. http://www.nyc.gov/html/planyc2030/downloads/pdf/planyc_progress_report_2010.pdf (accessed 05/10/2010).
5. Dwyer, J.F, H.W. Schroeder, and P.H. Gobster. 1991. The significance of urban trees and forests: toward a deeper understanding of values. Journal of Arboriculture 17(10):276-284.
6. Gieryn, T.F. 2000. A space for place in sociology. Annual Review of Sociology 26:463-497.
7. Gorman, J. 2004. Residents' opinions of the value of street trees depending on tree location. Journal of Arboriculture 30(1):36-44.
8. Grove, J.M., J. O'Neil-Dunne, K. Pelletier, D. Nowak, and J. Walton. 2006. A report on New York City's present and possible urban tree canopy: Prepared for Fiona Watt, Chief of the Division of Forestry and Horticulture. New York Department of Parks and Recreation, USDA Forest Service, Northern Research Station. 28 pp. http://nrs.fs.fed.us/nyc/local-resources/downloads/Grove_UTC_NYC_FINAL.pdf (accessed 10/01/2010).
9. Lewis, C.A. 1996. Green Nature/Human Nature: The Meaning of Plants in Our Lives. University of Illinois Press, Chicago. 148 pp.
10. Loukaitou-Sideris, A. and R. Ehrenfeucht. 2009. Sidewalks: Conflict and Negotiation over Public Space. MIT Press, Cambridge Mass. 328 pp.
11. Low, S.M. and I. Altman. 1992. Place attachment: a conceptual inquiry, pp. 1-12. In Low, S.M. and I. Altman, (Eds.). Place Attachment. Plenum Press, New York.
12. Kaplan, R. and S. Kaplan. 1989. The Experience of Nature: a Psychological Perspective. Cambridge University Press, New York. 340 pp.
13. Kaye, J.S, R. Polechronis, A. Reddy, and J. Rebold. 2009, February. City sidewalks trees and the law. New York State Bar Association Journal 24-27.
14. Kou, F.E. 2003. The role of arboriculture in a healthy social ecology. Journal of Arboriculture 29(3):148-154
15. Kou, F.E, M. Bacaicoa, and W.C. Sullivan. 1998. Transforming inner city landscapes: trees, sense of safety, and preference. Environment and Behavior 30(1):28-59.
16. Manzo, L. 2005. For better or worse: exploring multiple dimensions of place meaning. Journal of Environmental Psychology 25(1):67-86.
17. McIntyre, L. 2008, February. Shared wisdom. Landscape Architecture 88-93.
18. McPherson, E.G., J.R. Simpson, P.J. Peper, S.L. Gardner, K.E. Vargas, and Q. Xiao. 2007. Northeast Community Tree Guide: Benefits, Costs and Strategic Planting. General Technical Report, PSW GTR-202. Albany, CA: U.S. Department of Agriculture, Forest Service, Pacific Southwest Research Station. 106 pp.
19. New York City Department of Transportation. 2008. Sidewalks, the New York City Guide for Property Owners. Sidewalks and Inspection Management, Office of Sidewalk Management, New York City. http://www.nyc.gov/html/dot/html/faqs/sidewalkfaqs.shtml (accessed 05/13/2010)

20. Nowak, D.J., R.E. Hoehn, D.E. Crane, J.C. Stevens, and J.T. Walton. 2007. Assessing urban forest effects and values, New York City's urban forest. Resource Bulletin NRS-9. Newtown Square, PA, US Department of Agriculture, Forest Service, Northern Research Station. 22 pp.
21. Peper, P.J., E.G. McPherson, J.R. Simpson, S.L. Gardner, K.E. Vargas, and Q. Xiao. 2007, April. New York City, New York, Municipal Forest Resource Analysis. Technical Report to Adrian Benepe, Commissioner, Department of Parks & Recreation, New York City, New York. Center for Urban Forest Research, USDA Forest Service, Pacific Southwest Research Station. 65 pp.
22. Proshansky, H.M., A.K. Fabian, and R. Kaminoff. 1983. Place identity: physical world socialization of the self. Journal of Environmental Psychology 3:57-83.
23. Sack, R.D. 1986. Human Territoriality: Its Theory and History. Cambridge Studies in Historical Geography. Cambridge University Press, Cambridge, UK. 256 pp.
24. Schroeder, H., J. Flannigan, and R. Coles. 2006. Residents' attitudes toward street trees in the UK and US communities. Arboriculture and Urban Forestry 32(5):236-246.
25. Stephens, M. 2008, January/February. A greater and greener New York City. City Trees, Journal of the Society of Municipal Arborists. 22-23,38.
26. Sommer, R. 2004. Territoriality, pp 541-544. In Spielberger, C. D. (Ed). Encyclopedia of Applied Psychology, Volume 3. Elvsevier, California.
27. ------. 2003. Trees and human identity, pp 179-204. In Clayton, S. and S. Opotow. (Eds.). Identity and the Natural Environment: The Psychological Significance of Nature.
28. Sommer, R., F. Learey, J. Summit and M. Tirrell. 1994. The social benefits of resident involvement in tree planting. Journal of Arboriculture 20(3):170-175.
29. Summit, J. and E.G. McPherson. 1998, March. Residential tree planting and care: a study of attitudes and behavior in Sacramento, California. Journal of Arboriculture 24(2): 89-96.
30. Summit, J. and R. Sommer. 1998. Urban tree-planting programs - a model for encouraging environmentally protective behavior. Atmospheric Environment 32(1):1-5.
31. Svendsen, E.S. and L. Campbell. 2005. Living Memorials Project: Year 1 Social and Site Assessment. US Forest Service General Technical Report NE-333.
32. Strauss, A. and J. Corbin. 1990. Basics of Qualitative Research: Grounded Theory Procedures and Techniques. Sage Publications, California. 270 pp.
33. Taylor, R.B. 1988. Human territorial functioning: An empirical, evolutionary perspective on individual and small group territorial cognitions, behaviors, and consequences. Cambridge University Press, Cambridge, UK.
34. Tidball, K.G., M.E. Krasny, E.S. Svendsen, L. Campbell, and K. Helphand. 2010. Stewardship, Learning, and Memory in Disaster Resilience. Environmental Education Research 16(5):341-357.
35. Ward, K.T. and Johnson, G.R. 2007. Geospatial methods provide timely and comprehensive urban forest information. Urban Forestry & Urban Greening 6(1):15-22.
36. Wickham, T.D. and H.C. Zinn. 2001. Human territoriality: an examination of a construct, pp 35-39. In Kyle, G. (Ed.). Proceedings of the 2000 Northeastern Recreation

Research Symposium. General Technical Report NE-276. Newtown Square, PA: U.S. Department of Agriculture, Forest Service, Northeastern Research Station.

37. Westphal, L.M. 2003. Urban greening and social benefits: a study of empowerment outcomes. Journal of Arboriculture 29(3):137-147.
38. Williams, K. 2002. Exploring resident preferences for street trees in Melbourne, Australia. Journal of Arboriculture 28(4):161-170.
39. Wolf, K.L. 2003. Introduction to urban and community forestry programs in the United States. Landscape Planning and Horticulture 4(3):19-28.
40. ------. 2008. With plants in mind: social benefits of civic nature. Master Gardener 2(1):7-11.

CHAPTER 5

It's Not Easy Going Green: Obstacles to Tree-Planting Programs in East Baltimore

MICHAEL BATTAGLIA, GEOFFREY L. BUCKLEY, MICHAEL GALVIN, AND MORGAN GROVE

5.1 INTRODUCTION

While the urban forest is valued for the many environmental benefits it provides—such as reducing storm water flow, impeding soil erosion, and mitigating the urban heat island effect—a large and growing body of evidence points to the social and public health benefits of strategically planted trees. These include improvements to human health (Takano et al. 2002; Lovasi et al. 2008; Mitchell and Popham 2008), energy savings (Akbari and Konopacki 2005), and higher market values for homes (Payton et al. 2008; Sander et al. 2010). An increase in urban tree canopy (UTC) has also been linked to lower crime rates (Kuo and Sullivan 2001; Troy et al. 2012). For these and other reasons, cities across the U.S. are measuring tree canopy, adopting UTC goals, and developing programs to pursue these goals (United States Conference of Mayors 2008). Grow Boston

Greener, Million Trees LA, MillionTreesNYC, and The Chicago Tree Initiative are just a few examples of programs with ambitious plans in place to increase canopy coverage in their respective cities.

Given the challenges of growing trees in an urban environment, advocates acknowledge that only a mix of planting on public and privately-owned and managed lands will allow cities to achieve a broad range of UTC goals (Grove et al. 2006). Thus, cities like New York have adopted an "All Lands, All People" approach, which takes into consideration the tree-growing potential of all urban lands—from parks and public rights-of-way to residential parcels, commercial properties, and vacant lots. This approach embraces cooperation and collaboration among government agencies and NGOs, and promotes the collection and integration of social and ecological information (Locke et al. 2013).

To promote expansion of the UTC as well as safeguard a city's investment in trees, Grove et al. (2006), Raciti et al. (2006), and Locke et al. (2010) recommend adoption of a strategy that incorporates the "Three P's"—Possible UTC, Preferable UTC, and Potential UTC. The first step involves mapping Possible UTC. Possible UTC refers to any non-road, nonbuilding, or non-water land; that is, any location in the city where it is biophysically possible to plant trees. As living components of the urban ecosystem, trees must be planted in locations—and under conditions—that permit their survival. This may be difficult in an urban environment that lacks open space. The second step is to determine Preferable UTC; that is, identify where it is socially desirable to plant trees. In essence, where are trees needed and where are they wanted? This stage opens the door to public involvement in the decision-making process. Finally, Potential UTC centers on the economic feasibility of planting trees in a given location.

Like many cities, Baltimore is seeking to expand its urban tree canopy. In 2006, government officials launched TreeBaltimore, an initiative to double the city's tree canopy to 40 percent by 2036. Although overall coverage has increased since implementation of the new urban forest management plan, many parts of the city still have extremely low canopy cover (Galvin et al. 2006; O'Neil-Dunne 2009). To ensure that all citizens have access to the benefits of urban trees, it is imperative that resource managers and other decision makers recognize and address these disparities—a concern driving research agendas in many U.S. cities (e.g., Landry and Chakraborty 2009; Danford et al. 2014; Wolch, Byrne, and Newell 2014).

In this paper we address several issues associated with the Possible and Preferable components of a city's urban tree program and its ability to achieve a UTC goal. Our research focuses on two neighborhoods in Baltimore: Madison-Eastend and Berea. These two neighborhoods are high priority areas for increasing UTC (Locke et al. 2013) and have a history of unsuccessful tree planting programs since the 1960s.

We explore three research questions. First, is there sufficient space in the Madison-Eastend and Berea neighborhoods of East Baltimore to support an aggressive tree planting effort? Second, do residents in these two districts want more trees and, if so, are they willing to support tree-planting programs? Finally, we ask whether a change in the ethnic profile of these two neighborhoods since the 1960s has caused a shift in the way trees are perceived. Ultimately, a goal of this research is to help urban forestry personnel more effectively manage the city's urban forest by better understanding some of the variation in perceptions, values, and preferences for urban trees among urban residents.

5.2 PERCEPTIONS OF THE URBAN FOREST

While the benefits and costs of urban trees are well documented, less is known about the complex relationship that exists between people and urban green spaces (Balram and Dragicevic 2005). More specifically, how do residents of different cultural and socioeconomic backgrounds perceive and value the urban forest? The question is a significant one as failure to address the needs and desires of residents can pose problems for resource managers pursuing UTC goals. This is especially true if the city in question must depend on citizen support and cooperation to ensure the survival of young trees (Lu et al. 2010).

An early survey conducted in Detroit found that 63 percent of residents preferred to live in neighborhoods where the streets were lined with shade trees and small flowering trees. Only two percent responded that they did not want trees on their streets. The benefits identified most often by respondents were "pleasant to look at," "gives shade," and "increases property values." The participants were 70 percent African American and 30 percent white, with a relatively even distribution of income levels (Getz et al. 1982). A study carried out in a suburb of New Orleans produced similar results, with "aesthetic/visual," "gives shade," and "attracts wildlife" emerging as the most important perceived benefits. Eighty-six percent of respondents said that protecting trees was highly important, with 80 percent saying they would pay higher taxes to maintain the urban forest (Lorenzo et al. 2000).

Lohr et al. (2004) administered a nationwide phone survey to identify both perceived benefits and perceived problems relating to urban trees. According to the survey, the most important reasons to have trees were to "shade and cool" and "help people feel calmer." When asked about problems associated with trees, residents mentioned allergies and obstruction of store signs. The authors also determined that older respondents and those with higher levels of educational attainment were more likely to link trees with quality of life. Gorman's (2004) survey results from State College, Pennsylvania also suggest a correlation between positive attitudes toward urban trees and higher levels of educational attainment. Respondents in this study listed "give shade," "pleasing to the eye," "flowers on tree," "neighborhood more livable," and "increase property value" as positive attributes of trees. Negative features related to public safety, such as damage to sidewalks and power lines. In their study of Alabama's urban forests, Zhang et al. (2007) found that awareness of forestry programs, employment, age (in this case, 56 years or younger), and annual income ($75,000 and higher) correlated positively with willingness to contribute money and volunteer time to urban forestry activities. Race, gender, and residence were not significant factors when it came to explaining attitudes toward urban trees (Talbot and Kaplan 1984).

Preferences for open space and recreation areas are often discussed in the context of culture (Gobster 2002; Elmendorf et al. 2005; Pincetl and Gearin 2005). In such cases, "culture" is inferred through race or ethnicity. Fraser and Kenney (2000), for example, reported that tree preferences in Toronto, Canada were divided along ethnic lines. Their findings indicate that residents of English descent prefer large shade trees, while Portuguese and Italian residents favor small fruit-bearing trees. Meanwhile, Chinese residents did not encourage tree planting in their neighborhoods.

The authors maintain that these preferences are intimately tied to the landscape histories of each respective group's country of origin. Similar to Lohr et al. (2004), who found that a significantly lower percentage of African Americans and Asian Americans said trees were important to quality of life compared to other ethnicities, several studies suggest that African Americans tend to favor parks and recreational areas with fewer trees due to concerns about safety and crime (e.g., Gobster 2002; Brownlow 2006; Lewis and Hendricks 2006).

5.3 STUDY AREA

East Baltimore is one section of the city that has long exhibited a noticeable lack of trees. In an early attempt to increase UTC, the mayor's office, in 1965, allocated $326,000 to plant 8,000 street trees per year over a multi-year period. However, a tree survey conducted by city forester Fred Graves revealed that the cost of planting trees in East Baltimore alone—one of fourteen city sections surveyed—would exceed $385,000, more than the entire budget for the tree-planting effort and more than four times higher than the next most costly section of the city. Graves noted that East Baltimore was "practically denuded of trees" and that "the entire area has solid cement sidewalks without openings for trees" (quoted in Buckley 2010, 170). Despite high cost estimates, the Division of Forestry started to plant trees in East Baltimore two years later. It was at this time that city officials discovered another problem: many residents opposed tree-planting programs in their neighborhoods. Known in the local press as the city's "tree rebels," these residents claimed to prefer "clean, uncluttered concrete" to urban trees (Figure 1). They further argued that, "Trees belong in the country, not the city." According to Graves, this anti-tree sentiment was not evident in other parts of the city (quoted in Buckley 2010, 171-172).

Much has changed in the fifty years since residents of East Baltimore voiced opposition to the city's plans for tree planting. As manufacturing jobs declined, so too did East Baltimore's population. Formerly occupied by a diverse mix of immigrants from southern and eastern Europe, the area is now inhabited largely by African Americans. One thing remains

FIGURE 1: An example of "clean, uncluttered concrete" in Baltimore ca. 1948. Note the lack of tree pits in this block and the attention given to the condition of the marble steps (Photo taken by A. Aubrey Bodine, Courtesy of the Maryland Historical Society).

constant—the area lacks trees, and thus it is an important target area for TreeBaltimore. However, the decision to plant trees here should not be made hastily. The limited budget of the Division of Forestry—and programs like TreeBaltimore—makes site selection extremely important. Many variables must be taken into consideration to ensure that new tree planting will be successful.

According to Galvin et al. (2006) and O'Neil-Dunne (2009), Baltimore will not be able to meet its UTC goal of 40 percent coverage by planting trees only in parks and along streets. In fact, such a strategy, even if carried out to its maximum potential, would fall far short. The greatest opportunities for increasing tree canopy in Baltimore depend on other lands. Moreover, they depend on the cooperation and collaboration of private landowners and other community stakeholders all across Baltimore.

Two neighborhoods, Madison-Eastend and Berea, were selected as study areas for this research (Figure 2). Madison-Eastend is the smaller of the two, occupying 66.7 acres just north of Patterson Park. Berea, meanwhile, comprises an area of 217.6 acres including the expansive Baltimore Cemetery. Selection was based on several criteria. First, it was necessary to pick neighborhoods near the "tree rebel" area of the 1960s to gauge how attitudes toward tree planting may have changed with time and shifting demographics. Another important criterion was to select neighborhoods with differing physical characteristics. Madison-Eastend and Berea vary greatly when it comes to lot and house size, as well as available green space, allowing us to investigate plantable space and resident preferences in different contexts. Finally, the selection was based on a tree planting prioritization scheme developed for Baltimore's neighborhoods. Modeled after Nowak et al. (2007), the plan assigned each neighborhood an index score between 1 and 100. The index is based on population and tree cover densities, with a score of 100 indicating a high population density and low percentage canopy cover and a score of 1 indicating a low population density and high percentage canopy cover. Of the 271 neighborhoods with adequate data, Madison-Eastend ranked tenth and Berea twenty-third when it came to greatest need for tree planting (Battaglia 2010).

With respect to physical differences, Madison-Eastend is characterized by high-density row houses and a noticeable lack of greenery. Built between 1890 and 1920, the houses are situated close to the street with no

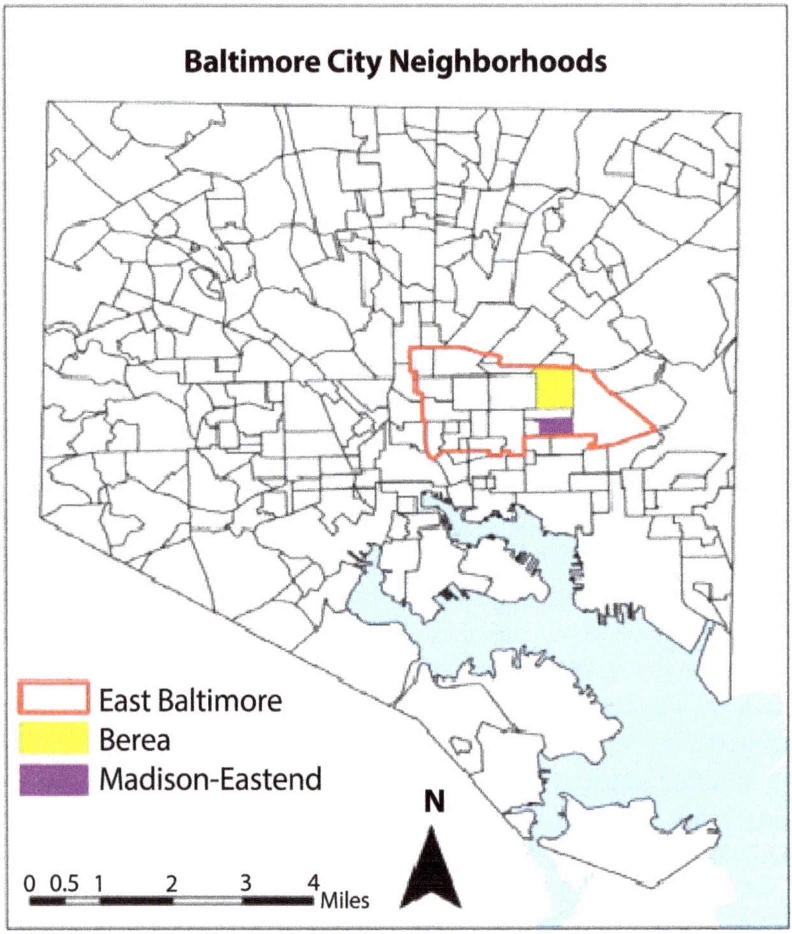

FIGURE 2: Berea and Madison-Eastend correspond roughly with the location of East Baltimore's "tree rebels" of the 1960s.

space for front yards. Most have a small paved lot in the back, which connects to an alley. Much of the area's green space is concentrated at Bocek Park in the northeast corner of the neighborhood, and in the front of the office buildings located nearby. In contrast, Berea's row houses were constructed later, are relatively large, and have both a front and a backyard.

Most residential streets are lined by areas of grass between the sidewalk and the street, known as "tree strips" or "tree lawns." Two neighborhood elementary schools and several churches contain additional green space.

Regarding the area's demographic makeup, significant change has occurred in East Baltimore over the last several decades. Between 1970 and 2010 in Madison-Eastend, an area once dominated by working class immigrants of European descent, the African American population increased dramatically, from 14.24 percent to 90.26 percent. At 96.30 percent, Berea's African American population, having secured a foothold in the neighborhood much earlier, has remained relatively constant over the same period. Citywide, African Americans today make up approximately 63.82 percent of the total population (BNIA 2013). Both areas experienced an overall decline in total population from 1970 to 2010.

5.4 METHODS AND FINDINGS

5.4.1 POSSIBLE UTC

For the purposes of this study, plantable area refers to any pervious surface not covered by tree canopy. To assess plantable area (Possible UTC) within Madison-Eastend and Berea, geo-spatial analyses were conducted using ArcGIS 9.3 software. GIS shape files of neighborhood boundaries, parcel boundaries, street centerlines, building footprints, pavement edge, tree canopy, and other planimetric data, along with 2008 aerial imagery, were obtained from the Mayor's Office of Information Technology (MOIT). An overlay method, similar to previous studies (Grove et al. 2005; Raciti et al. 2006), was combined with tree pit data we collected to produce final plantable area maps for each neighborhood.

Both neighborhoods in the study area have a considerable amount of possible tree planting space relative to the neighborhoods' size. The plantable area totals 23.55 acres for Berea and 7.08 acres for Madison-Eastend (Figures 3 and 4). Residential plantable space comprises a significant portion of the total for Berea—especially the eastern section—but only a very small amount for Madison-Eastend. This is because many of Berea's dwellings have both front and backyards. Both neighborhoods have

FIGURE 3: Berea Plantable Space. Many of the houses in Berea are set back from the street and possess both front and backyards. As a result, there is a great deal of residential plantable space. There are also opportunities to plant trees along public rights-of-way (PROW) and on properties owned by schools, churches, businesses, and the City of Baltimore.

planting opportunities along public rights-of-way (PROW) which include all land area that is not part of a parcel, such as roads, alleys, sidewalks, and other public transportation corridors. Other plantable space includes parcels managed by the City of Baltimore, schools, church groups,

FIGURE 4: Madison-Eastend Plantable Space. Unlike Berea, there is comparatively little residential plantable space in Madison-Eastend. This is due to differences in housing type and lot size. However, there are opportunities to plant trees along public rights-of-way (PROW) and in "other" plantable spaces, especially Bocek Park, which occupies the eastern third of the neighborhood.

businesses, or other private entities. Both neighborhoods possess significant plantable space under this category.

With respect to planting opportunities along PROW, we counted 224 street trees and a total of 13,881 meters of potential planting length along Berea's streets, not including the cemetery, alleys, or streets with sidewalks less than four feet wide. According to these numbers, there is one street tree for every 61.87 meters of roadside length. With 7.62 meters between pits, there is a potential to add many more trees. However, the best possibility for planting along the public rights-of-way in Berea is on the open tree lawns. The sum of the open tree lawns' lengths is 2,972.7 meters. Most would be suitable for small or medium-sized trees. (Note: Although data specifying the locations of underground cables and sewage lines were not available, we eliminated from consideration sites where obstacles to tree planting were clearly present, such as locations with overhead electrical wires and street lights.) If tree planting along the public rightsof- way were prioritized, 390 new trees could be installed along the tree lawns alone. Combined with planting in empty tree pits, there is an opportunity to plant 418 street trees in Berea, which would almost triple the number in the neighborhood to 642.

Madison-Eastend has 83 street trees and a total of 6,948.5 meters of space along its roads, yielding an average of one street tree per 83.5 meters throughout the neighborhood. If the goal were to maximize street tree planting, at least 10 trees could be planted along a corridor that currently accommodates just one. The amount of actual plantable space along the public rights-of-way is considerably less. Because of the type of row houses present in Madison-Eastend, the length of the open tree lawns is only 194 meters. At 1.22 meters wide they are able to accommodate small or medium-sized trees. If planting were maximized, 26 new trees could be planted. If every location along Madison-Eastend's public rights-of-way were planted, the number of street trees could be increased from 83 to 140. That said, if both neighborhoods were to maximize tree planting they could increase existing canopy cover significantly—from approximately 5.26 to 16.08 percent in Berea, and from 6.23 to 16.84 percent in Madison-Eastend (Table 1). Cumming et al. (2001) found a stocking level of 13.9 percent for roadside trees across the state of Maryland. Using their 15.24 meter spacing, stocking levels in Berea and Madison-Eastend would be somewhat higher at 24.6 percent and 18.2 percent, respectively.

Table 1. Existing and possible tree canopy cover in the study area.

	Berea	Madison-Eastend
Total Area (acres)	217.61	66.74
Tree Canopy (acres)	11.44	4.16
Tree Canopy (%)	5.26	6.23
Plantable (acres)	23.55	7.08
Plantable (%)	10.82	10.61
Possible Tree canopy (acres)	34.99	11.24
Possible Tree canopy (%)	16.08	16.84

5.4.2 PREFERABLE UTC

The measures of Possible UTC only take into account where it is biophysically possible to plant trees. The next step was to understand preferences for UTC. Thus, we sought to explore how trees were perceived and valued in the study area and whether residents wanted and would care for additional trees. To determine this, we interviewed residents in both Madison-Eastend and Berea. Initial contacts with interview subjects were facilitated by the TreeBaltimore coordinator; additional respondents were contacted by referral or during the pit survey. After Institutional Review Board (IRB) approval was granted, research trips to Baltimore were carried out in December of 2009 and April 2010. In total, 26 interviews were conducted. Sixteen of the residents lived in Berea, while ten resided in Madison-Eastend. Sixteen of the respondents were male. All of the residents interviewed were African American and all were at least 18 years of age. A semi-structured interview style was adopted to allow flexibility in the event an informant wished to speak about a topic not covered by the interview guide. Most interviews took approximately 15 minutes to complete. Interview notes were transcribed and later coded. The coding was analytic in nature with each interview assigned codes based on the nature of subjects covered.

While the interview data do not express the views of everyone in the study area, they provide a wealth of information regarding how some residents understand trees. There were several who said they think tree planting is a good idea, citing many of the same benefits mentioned in earlier studies. Some of these, such as aesthetic enhancement and shade provision, were widely acknowledged in Madison-Eastend and Berea. Less obvious benefits, such as water quality improvement and carbon sequestration, were referenced only a few times. While some residents expressed support for new planting, others opposed it. Their reasons for wanting to limit tree planting were wide ranging and included items not mentioned in earlier surveys. In the following two sections, we summarize our findings in terms of residents' positive and negative perceptions.

5.4.2.1 POSITIVE PERCEPTIONS

Of the 26 interviews conducted in the study area, 14 revealed some type of positive perception of urban trees. One of the most widely understood positive attributes of trees was their ability to provide shade. Baltimore has a humid subtropical climate, with temperatures sometimes reaching 100 degrees Fahrenheit during summer months (National Oceanic and Atmospheric Administration 2013). These high temperatures can cause discomfort for residents, especially those whose homes are not air-conditioned, a point confirmed by a male resident of Madison-Eastend: "Man, it sure does get hot here, if you're around in summertime you see everybody sitting outside. No one wants to stay cooped up inside in the heat. Some trees would be real nice to have, especially some big shady ones. Maybe a nice big one right in front of my house!" This feeling was widespread among interview participants. It was especially important to those who did not have access to shaded outdoor areas in Madison-Eastend. One woman stated, "It's like sitting on top of a stove . . . out here." Several interviewees mentioned the common summertime practice of relaxing on the front stoop or porch. This was observed during the summertime tree surveys we conducted, when the sidewalks of Madison-Eastend filled with people during the mid to late afternoon hours. It was also clear that people gravitated to the side of the street that was not in direct sunlight.

Berea residents appreciated shade as well. One woman remarked, "Our block is lucky, we have all these big trees, they keep us cool when it gets to be summer. I know a lot of these blocks don't have any trees at all." A recent high school graduate commented that he was aware of the urban heat island effect, and that he knew tree shade would help reduce it by lowering temperatures. Some residents said they understood that strategically planted trees could save them money on their energy bills. A man in Berea remarked that he was aware of reductions in energy costs through tree planting and that he had planted a tree in his backyard the previous summer for that reason. Another stated that he knew shade trees could reduce energy costs and, further, that he would like to plant a tree but his yard in Berea was too small.

Residents also valued the aesthetic appeal of trees. A woman living in Berea said, "This block just looks better, people here plant flowers and you get the flowers and the trees all together and it looks nicer than some of the other blocks around here." Another woman added, "I'm glad I live here. It's not the best part of the city, we have our problems, you know? But compared to some other parts, like across the tracks down there, they got it bad. You can go for blocks before you see a tree. . . . That's just depressing." A female resident of Madison-Eastend concurred: "Beautification is important in this area. It's a rough neighborhood. I think if you make it look nicer it wouldn't feel as rough."

The possibility of trees contributing to the mitigation of global climate change was mentioned on two occasions. Although the effectiveness of urban trees as pollution filters and greenhouse-gas reducers has been questioned (Nowak et al. 2007; Pataki et al. 2011), both interviewees had strong opinions on the subject. "I know all about global warming, we need to plant trees to stop it. I get that. I do know they provide oxygen. They take the bad stuff out of the air." The other respondent felt that it was one of the most important characteristics of urban trees. "We need more oxygen in our environment. Without oxygen, there can be no life. . . . So we have two choices, learn to treat our earth right, stop chopping down the rainforest, or start looking for another planet to inhabit. It starts right here though. Planting trees is very important."

5.4.2.2 NEGATIVE PERCEPTIONS

Although the ability to attract wildlife is often listed as a benefit of the urban forest (Dwyer et al. 1992; McPherson et al. 1997), none of the interview participants in East Baltimore viewed wildlife in this way. Instead, animals, such as birds, were considered nuisances. Bird droppings, in particular, were a source of frustration for residents. An elderly woman who has lived in Berea for over 40 years stated: “We have enough trees. We don’t need any more. We got two on this block, and that’s more than enough. I don’t think most folks want trees. Everybody always complains about the bird manure anyway.” Even those who otherwise were in favor of tree planting mentioned birds as a problem.

Insects were another perceived problem. Echoing the sentiments of an anti-tree rebel from the 1960s (Buckley 2010), a resident of Berea was not happy about a recent spike in the population of “caterpillars.” Several participants also mentioned rats as a reason for opposing new planting, two of whom were convinced that trees attracted rats. According to 2009 figures from the Baltimore Neighborhood Indicators Alliance, the number of reported incidents of rats per 1000 residents was 215.70 for Madison-Eastend and 118.44 for the Clifton-Berea community statistical area (CSA). The citywide average was 59.69 (BNIA 2012).

Several residents said they were allergic to tree pollen. A resident of Madison-Eastend stated that after growing up in a part of Washington, D.C. that had many trees he was relieved to not have as many problems. In his words, “I don’t want to have to start taking allergy pills again.” A resident of Berea who otherwise supported aggressive tree planting lamented that he has been dealing with allergy-induced asthma his whole life, but that it was a necessary tradeoff.

Property damage from trees was another issue residents discussed. When initially asked how she felt about tree planting programs, an elderly woman in Berea responded “No thank you! No trees for me!” Throughout her time as a homeowner in East Baltimore she has had numerous problems with tree roots breaking her water pipes. A Madison-Eastend man pointed to a group of vacant row houses along Glover Street all of which had been infiltrated by tree branches. He maintained that these trees caused damage to electrical wires, and that he had experienced several power outages in the previous year. Heynen et al. (2006) described a similar situation in an African American neighborhood in Milwaukee, where trees were often removed due to property damage.

While none of the interview participants admitted to a preference for "clean, uncluttered concrete" like the former inhabitants of East Baltimore, many found certain aspects of the urban forest displeasing. In particular, residents did not appreciate the dead trees. When asked how he felt about additional tree planting on his street, a Berea resident responded, "Why would I want another tree when I can't get rid of this dead one? I've been on the city for a year to get rid of it but it's still right there." Another resident of Berea added: "I have lived here for over 20 years now, and I have seen trees get planted. Those trees that get planted just die. . . . The city wants to plant more trees, why the hell don't they just take care of the ones already here?" Another man agreed: "Sure, I think planting trees is a good idea. It's also a good idea for them to take down the dead ones before they start planting more." Several interviewees worried about large dead trees or limbs falling onto their houses during storms. The persistence of dead trees in the urban landscape and the high mortality that can result from lack of community support has long been a concern of residents and resource managers (Sklar and Ames 1985; Roman et al. 2013).

As with many large urban areas, parts of Baltimore have significant drug problems. According to 2011 statistics compiled by the Baltimore Neighborhood Indicators Alliance, juvenile drug-related arrests per 1000 people have declined over the last five years in the Clifton- Berea CSA and Madison-Eastend. Nevertheless, at 63.7 and 49.07, respectively, they remain significantly higher than the Baltimore City average of 30.26 (BNIA 2013). Interview participants linked trees to the drug trade on several occasions. One respondent from Madison- Eastend said, "No man, no, we don't need more trees. That's just another place to hide drugs. We don't need more of that around here." Another remarked "When I was a younger man running around on the streets, we used to use them (trees) as a drop spot (for drugs)."

Some neighborhood members worried that tree planting would be carried out for the wrong reasons. One resident of Berea was skeptical of outsider interference in his community. : His mistrust of outsiders stems from the recent bulldozing of entire blocks just a short walk to the west to make room for an expansion of Johns Hopkins Hospital. In particular, he worried about how other development plans might affect his community in the future. In his opinion, tree planting would be followed by gentrification and displacement of the remaining population of Madison-Eastend.

As Atkinson (2003) notes, gentrification rarely benefits underserved communities, leading Wolch, Byrne, and Newell (2014) to recommend neighborhood greening initiatives that are "just green enough" to improve the lives of residents but not enough to trigger sharp increases in property values.

Another resident of Berea was skeptical of urban trees for a different reason. He was concerned that tree planting was just the "flavor of the week" for whatever politician was trying to get elected to office. When asked about the possibility of trees on his street he recounted previous city initiatives that proved ineffectual. His feelings reflect the recent political turmoil in the city, where political corruption led to the resignation of the mayor (Bykowicz 2010).

Many citizens argue that there are more pressing problems that need to be addressed before the city dedicates funds to tree planting. One man suggested spending money on trash cans. Another questioned why the city had cut its trash collection days but was willing to spend more on trees. A woman from Madison-Eastend remarked, "It's just dirty around here. There's trash everywhere, people don't care." According to the BNIA there were 267.7 reports of dirty streets and alleys per 1000 people in Madison-Eastend in 2011, the highest rate in the city. The Clifton-Berea CSA ranked fifth highest with 171.87 reports per 1000 people (BNIA 2013).

Before trees are planted on or adjacent to a residence, homeowners must sign a waiver agreeing to water young trees and take basic steps to ensure their survival (TreeBaltimore 2007). Therefore, in addition to answering questions about their perceptions of trees, residents were asked how receptive they thought their community would be to tree planting initiatives. When asked whether he thought his neighbors would care for trees, the recent high school graduate from Berea stated: "It all depends. I think on this block it would work. I'd water a tree!

Some of these houses around here though, well I don't know (laughs). Some people really don't care about that type of thing." An elderly gentleman in Berea was less optimistic. "More trees would be nice, but we have already had trees on this block and they die. People don't water them. And most of the time, even if they do get watered, they get killed anyway by the children. The children around here have no respect for anything. They run wild and do what they want. I've seen them tear little trees apart."

A woman from Berea differentiated between homeowners and those who rent, indicating she was not confident renters would put in the effort to take care of newly planted trees: "Most of us around here own our homes. Most of us have lived here for a long time. We care about our neighborhood and the way it looks, obviously. You go down that way (pointing south) though, I don't think they're gonna help out too much. They mostly rent. Folks that rent, why should they care?" Her belief that renters are less enthusiastic about tree planting programs is supported by the literature. In their study of Milwaukee's urban forest, Perkins et al. (2004) discovered that only 11 percent of those who took advantage of a free tree-planting program were renters in a city where 55 percent of homes are occupied by renters. In the study area, a majority of home occupants are renters as well. In Clifton-Berea, just 34.35 percent of housing units were owneroccupied in 2011 (BNIA 2013).

Older interview participants in Berea indicated that they thought attempts to plant trees in the future would fail because of changes in the population. According to several interviewees, a majority of the original African American residents of East Baltimore had migrated from rural parts of the South and knew how to take care of trees and plants. Now, only a few of the original transplants remain. The ability and desire to care for the natural world, they claim, has diminished among the people who have grown up in the city. As a former South Carolina resident living in Berea put it, "Some of these people don't know the difference between a pine tree and an apple tree." As Ziederman (2006) points out, the migration of African Americans from the rural south to the industrialized north brought not only people, but agricultural skills and preferences as well. For the aging residents of Madison-Eastend and Berea, trees may be representative of a landscape preference that is rapidly dying out.

5.5 CONCLUSIONS

In this paper we explored the potential for tree planting in two of East Baltimore's neighborhoods, Madison-Eastend and Berea. Fifty years after residents derailed a major treeplanting effort, and despite significant demographic changes, large sections of these neighborhoods still exhibit a noticeable lack of trees. In the 1950s and 1960s, the residents of East Baltimore, many of whom were immigrants from southern and eastern Europe, found urban trees socially undesirable. When Blacks from the American South arrived in increasing numbers after 1970, they likely brought with them different attitudes with respect to trees. While many may have viewed trees in a more positive light—perhaps even socially preferable—they inherited a landscape that was largely devoid of canopy cover during a period of disinvestment in America's cities. Trees may have been socially preferable, but the legacy of the area's former residents—virtually treeless neighborhoods—ensures that a major effort to increase UTC requires a significant economic investment (see also Boone et al. 2010 and Buckley et al. 2013). Today, a new generation of African American residents voice strong opinions both for and against tree planting in East Baltimore.

An important objective of our research was to determine whether a change in the ethnic profile of a community—in this case, from southern and eastern European to African American—might signal a change in the way trees are perceived. Recognizing the small sample size and limited geography of our exploratory research, the qualitative data presented here suggest that using ethnic groups as vehicles to make broad generalizations about the perceptions and preferences of many people is problematic. This result is supported by Li et al. (2007, 515), who argue that, "The cultural variability within purported ethnic groups may be as great, or greater, than the cultural variability between them." Failure to recognize variability within a cultural, racial, or ethnic group poses problems. At best, it leads to the perpetuation of stereotypes. At worst, it implies the acceptance of a form of environmental determinism. Our research indicates that most people's perceptions of trees were practical and developed through lived experiences.

With respect to Possible UTC, our research shows that there is room to increase tree canopy in the study area from approximately six percent to more than 16 percent, making Madison-Eastend and Berea prime targets for TreeBaltimore. In Berea, most of the plantable area is located on residential parcels. Although all of the homes are considered row houses, a majority of the homes in the eastern part of the neighborhood are on large parcels that include front and backyards. In contrast, Madison-Eastend has limited plantable area on residential land because the row houses lack front yards, and most backyards are paved. Bocek Park and the land along the neighborhood's northern border account for most of the plantable area. Although plantable area is limited along public rights-of-way, there is still considerable space for tree planting.

While measuring Possible UTC is an important first step, gauging the degree to which residents support tree planting in their neighborhoods (Preferable UTC) gives us a better indication of how successful investments in green infrastructure may prove in the long run. The interviews we conducted in Madison-Eastend and Berea reveal mixed attitudes towards trees. Fourteen of the 26 participants supported tree planting because of perceived benefits such as shade and beauty. However, several of these individuals expressed doubt that residents—especially those who rent—would maintain trees planted in front of their homes, supporting the argument that tree care can sometimes place an unacceptable burden on the shoulders of lower income residents (Landry and Chakraborty 2009). The 12 remaining participants opposed tree planting and discussed a variety of negative perceptions, often in great detail, ranging from problems with pests and allergies to concerns about gentrification and the management of existing trees.

A serious issue that civic leaders in Baltimore must address is how to handle residents' negative perceptions of trees. The academic community has clearly elucidated the many benefits provided by urban trees, and municipal policy in Baltimore and elsewhere reflects this enhanced understanding of the benefits of urban forests. This perspective is not shared by everyone, however, and the question of how to deal with it is a challenging one. Acknowledging residents' negative perceptions is necessary in order to move forward. Reminding residents of the many ways trees could benefit them may sway their opinions. However, any type of educational program in Madison-Eastend or Berea should be carefully formulated to address neighborhood conditions and concerns. Clearly, focusing on property value increases and attracting wildlife would deter some residents from supporting tree planting. Highlighting energy savings and mitigation of the urban heat island effect is more likely to make a favorable impression.

TreeBaltimore's challenge, then, is not simply to overcome the limitations of Possible UTC, but to enlist the support of residents and address their preferences and priorities. Two opportunities emerge from this study. The first opportunity relates to the management of older trees. As residents made clear in the interviews, there are deep-seated concerns regarding the maintenance of existing trees, including the removal of dead, dying, and hazardous trees. Finding a way to meet the needs of residents in this regard may help to generate support for future tree planting efforts. The second opportunity is related to citizen involvement in the decision-making process. Exploring new and innovative ways to engage and empower communities like Madison- Eastend and Berea offers resource managers a chance to both address negative attitudes toward urban trees and restore confidence in city government. Thus, while planting trees in disadvantaged neighborhoods like Madison-Eastend and Berea would help close the gap with respect to tree cover equity, it is also clear that city officials and resource managers also consider the care and health of urban trees over the long term and its effects on residents' perceptions, values, and preferences.

REFERENCES

1. Akbari, H., and S. Konopacki. 2005. Calculating energy-saving potentials of heat-island reduction strategies. Energy Policy 33(6):721-756.
2. Atkinson, R. 2003. Introduction: Misunderstood Saviour or Vengeful Wrecker? The Many Meanings and Problems of Gentrification. Urban Studies 40(12):2343–2350.
3. Balram, S., and S. Dragicevic. 2005. Attitudes toward urban green spaces: Integrating questionnaire survey and collaborative GIS techniques to improve attitude measurements. Landscape and Urban Planning 71:147-162.
4. Baltimore Neighborhood Indicators Alliance. 2012. http://www.bniajfi.org/vs/vital_signs/5. (accessed 5 January 2014).

5. Baltimore Neighborhood Indicators Alliance. 2013. Vital Signs 11. http://www.ubalt.edu/bnia/. (accessed 10 January 2014).
6. Battaglia, M. J. 2010. A Multi-Methods Approach to Determining Appropriate Locations for Tree Planting in Two of Baltimore's Tree-Poor Neighborhoods. M.A. Thesis. Ohio University, Athens, Ohio, USA.
7. Boone, C. G., M. Cadenasso, K. Schwartz, J. M. Grove, and G. L. Buckley. 2010. Landscape, vegetation characteristics, and group identity in an urban and suburban watershed: why the 60s matter. Urban Ecosystems 13(3):255-271.
8. Brownlow, A. 2006. Inherited fragmentations and narratives of environmental control in entrepreneurial Philadelphia. In In the Nature of Cities: Urban Political Ecology and the Politics of Urban Metabolism, eds. N. Heynen, M. Kaika, and E. Swyngedouw, 208-225. London: Routledge.
9. Buckley, G.L. 2010. America's Conservation Impulse: A Century of Saving Trees in the Old Line State. Chicago: Center for American Places and Columbia College.
10. Buckley, G.L., A.C. Whitmer, and J.M. Grove. 2013. Parks, Trees, and Environmental Justice: Field Notes from Washington, DC. Applied Environmental Education & Communication 12(3):148-162.
11. Bykowicz, J. 2010. Dixon Resigns. Baltimore Sun, 7 January.
12. Cumming, A.B., M.F. Galvin, R.J. Rabaglia, J.R. Cumming, and D.B. Twardus. 2001. Forest Health Monitoring Protocol Applied to Roadside Trees in Maryland. Journal of Arboriculture 27(3):126-138.
13. Dales, R.E., S. Cakmak, S. Judek, and F. Coates. 2008. Tree pollen and hospitalization for asthma in urban Canada. International Archives of Allergy and Immunology 146(3):241–247.
14. Danford, R.S., C. Cheng, M.W. Strohbach, R. Ryan, C. Nicolson, and P.S. Warren. 2014. What Does It Take to Achieve Equitable Urban Tree Canopy Distribution? A Boston Case Study. Cities and the Environment 7(1): Article 2. http://digitalcommons.lmu.edu/cate/vol7/iss1/2. (accessed 17 October 2014).
15. Dwyer, J.F., E.G. McPherson, H. Schroeder, and R.A. Rowntree. 1992. Assessing the Benefits and Costs of the Urban Forest. Journal of Arboriculture 18(5):227-234.
16. Elmendorf, W.F., F.K. Willits, and V. Sasidharan. 2005. Urban park and forest participation and landscape preference: A review of the relevant literature. Journal of Arboriculture 31(6):311-316.
17. Fraser, E.D.J. and W.A. Kenney. 2000. Cultural background and landscape history as factors affecting perceptions of the urban forest. Journal of Arboriculture 26(2):106-113.
18. Galvin, M.F., J.M. Grove, and J.P.M. O'Neil-Dunne. 2006. A Report on Baltimore's Present and Potential Urban Tree Canopy. Scenario. http://www.fs.fed.us/nrs/utc/reports/UTC_Report_BACI_2007.pdf. (accessed 17 October 2014).
19. Getz, D., A. Karow, J. Kielbaso. 1982. Inner city preferences for trees and urban forestry programs. Journal of Arboriculture 8(10):258-263.
20. Gobster, P.H. 2002. Managing urban parks for a racially and ethnically diverse clientele. Leisure Sciences 24:143-159.
21. Gorman, J. 2004. Residents' opinions on the value of street trees depending on tree allocation. Journal of Arboriculture 30(1):36-43.

22. Grove, J.M., W.R. Burch, and S.T.A. Pickett. 2005. Social Mosaics and Urban Community Forestry in Baltimore, Maryland. Introduction: Rationale of Urban Community Forestry. Continuities from Rural to Urban Community Forestry. In Communities and Forests: Where People Meet the Land, eds. R.G. Lee and D.R. Field, 248-273. Corvallis, OR : Oregon State University Press.
23. Grove, J.M., J. O'Neil-Dunne, K. Pelletier, D. Nowak, and J. Walton. 2006. A report on New York City's present and possible urban tree canopy: Prepared for Fiona Watt, Chief of the Division of Forestry and Horticulture. New York Department of Parks and Recreation, USDA Forest Service, Northern Research Station, USA.
24. Heynen, N., H.A. Perkins, and P. Roy. 2006. The political ecology of uneven urban green space: The impact of political economy on race and ethnicity in producing environmental inequality in Milwaukee. Urban Affairs Review 42(1):3-25.
25. Kielbaso, J.J. 1990. Trends and issues in city forests. Journal of Arboriculture 16:69-73.
26. Kuo, F. and W. Sullivan. 2001. Environment and Crime in the Inner City: Does Vegetation Reduce Crime? Environment and Behavior 33:343-366.
27. Landry, S.M., and J. Chakraborty. 2009. Street trees and equity: evaluating the spatial distribution of an urban amenity. Environment and Planning A 41:2651-2670.
28. Lewis, J.G., and R. Hendricks. 2006. A Brief History of African Americans and Forests(Unpublished collaboration between the Forest History Society and the USDA Forest Service).
29. Li, C., G. Chick, H. Zinn, J. Absher, and A. Graefe. 2007. Ethnicity as a variable in leisure research. Journal of Leisure Research 39:514–545.
30. Locke, D.H., J.M. Grove, W.T. Lu, A. Troy, J.P.M. O'Neil-Dunne, and B. Beck. 2010. Prioritizing preferable locations for increasing urban tree canopy in New York City. Cities and the Environment 3(1): article 4. http://escholarship.bc.edu/cate/vol3/iss1/4. (accessed 17 October 2014)
31. Locke, D.H., J.M. Grove, M. Galvin, J.P.M. O'Neil-Dunne, and C. Murphy. 2013. Applications of Urban Tree Canopy Assessment and Prioritization Tools: Supporting Collaborative Decision Making to Achieve Urban Sustainability Goals. Cities and the Environment 6(1): article 7. http://digitalcommons.lmu.edu/cate/vol6/iss7. (accessed 17 October 2014)
32. Lohr, V.I., C.H. Pearson-Mims, J. Tarnai, and D.A. Dillman. 2004. How urban residents rate and rank the benefits and problems associated with trees in cities. Journal of Arboriculture 30(1):28-34.
33. Lorenzo, A.B., C.A. Blanche, Y. Qi, and M.M. Guidry. 2000. Assessing residents' willingness to pay to preserve the community urban forest: A small-city case study. Journal of Arboriculture 26(6):319-324.
34. Lovasi, G.S., J.W. Quinn, K.M. Neckerman, M.S. Perzanowski, and A. Rundle. 2008. Children living in areas with more street trees have lower prevalence of asthma. Journal of Epidemiology and Community Health 62:647-649.
35. Lu, J.W.T., E.S. Svendsen, L.K. Campbell, J. Greenfield, J. Braden, K.L. King, and N. Falxa-Raymond 2010. Biological, Social, and Urban Design Factors Affecting Young Street Tree Mortality in New York City. Cities and the Environment 3(1): article 5. http://digitalcommons.lmu.edu/cgi/viewcontent.cgi?article=1069&context=cate. (accessed 17 October 2014)

36. McPherson, E.G., D. Nowak, G. Heisler, S. Grimmond, C. Souch, R. Grant, and R. Rowntree. 1997. Quantifying urban forest structure, function and value: the Chicago Urban Forest Climate Project. Urban Ecosystems 1:49–61.
37. Mitchell, R., and F. Popham. 2008. Effect of exposure to natural environment on health inequalities: An observational population study. Lancet 372:1655-1660.
38. National Oceanic and Atmospheric Administration. 2013. Baltimore/Washington International Airport Normals, Means, and Extremes. http://www.erh.noaa.gov/lwx/climate/bwi/NME.htm (accessed 12 December 2013).
39. Nowak, D.J. 1994. Urban forest structure: The state of Chicago's urban forest. In Chicago's Urban Forest Ecosystem: Results of the Chicago Urban Forest Climate Project. General Technical Report No. NE-186, eds. E. G. McPherson, D. J. Nowak, and R. A. Rowntree, 3-18. USDA Forest Service, Northeastern Forest Experiment Station.
40. Nowak, D.J., R.E. Hoehn III, D.E. Crane, J.C. Stevens, and J.T. Walton. 2007. Assessing urban forest effects and values, Washington, DC's urban forest. Resource Bulletin. NRS-1. USDA Forest Service, Northern Research Station.
41. O'Neil-Dunne, J.P.M. 2009. A Report on the City of Baltimore's Existing and Possible Tree Canopy. Burlington, VT: The Spatial Analysis Lab at the University of Vermont's Rubenstein School of the Environment and Natural Resources.
42. Pataki, D.E., M.M. Carreiro, J. Cherrier, N.E. Grulke, V. Jennings, S. Pincetl, R.V. Pouyat, T.H. Whitlow, and W.C. Zipperer. 2011. Coupling biogeochemical cycles in urban environments: ecosystem services, green solutions, and misconceptions. Front Ecol Environ 9(1):27–36.
43. Payton, S.B., G.H. Lindsey, J.R. Wilson, J.R. Ottensmann, and J.S. Man. 2008. Valuing the benefits of the urban forest: a spatial hedonic approach. Journal of Environmental Planning and Management 56(6):717-736.
44. Perkins, H., N. Heynen, and J. Wilson. 2004. Inequitable access to urban reforestation: the impact of urban political economy on housing tenure and urban forests. Cities 21(4): 291-299.
45. Pincetl, S., and E. Gearin. 2005. The Reinvention of Public Green Space. Urban Geography 26:365-384.
46. Raciti, S., M. F. Galvin, J. M. Grove, J. P. M. O'Neil-Dunne, A. Todd, and S. Clagett. 2006. Urban tree canopy goal setting: A guide for Chesapeake Bay communities. Annapolis, MD: USDA Forest Service, Northeastern State and Private Forestry, Chesapeake Bay Program Office. http://www.jmorgangrove.net/Morgan/UTCFOS_files/UTC_Guide_Final_DRAFT.pdf. (accesed 17 December 2006).
47. Roman, L.A. 2013. Urban Tree Mortality. Ph.D. Dissertation. University of California, Berkeley.
48. Sander, H., S. Polasky, and R.G. Haight. 2010. The value of urban tree cover: a hedonic property price model in Ramsey and Dakota counties, Minnesota, USA. Ecological Economics 69(8):1646-1656.
49. Sklar, F. and Ames, R.G. 1985. Staying Alive: Street Tree Survival in the Inner City. Journal of Urban Affairs 7:55-66.
50. Takano, T., K. Nakamura, and M. Watanabe. 2002. Urban residential environments and senior citizens' longevity in megacity areas: The importance of walkable green spaces. Journal of Epidemiology and Community Health 56(12):913-918.

51. Talbot, F. and R. Kaplan. 1984. Needs and Fears: the response to trees and nature in the inner city. Journal of Aboriculture 10(8): 222–228.
52. TreeBaltimore. 2007. Urban Forest Management Plan Draft. City of Baltimore. http://www.ci.baltimore.md.us/government/recnparks/downloads/TreeBaltimore%20Urban%20Forest%20Management%20Plan.pdf. (accessed 10 July 2008).
53. Troy, A., J.M. Grove, and J. O'Neill-Dunne. 2012. The relationship between tree canopy and crime rates across an urban-rural gradient in the greater Baltimore region. Landscape and Urban Planning 106:262–270.
54. United States Conference of Mayors. 2008. Protecting and Developing the Urban Tree Canopy: A 135-City Survey. Prepared by City Policy Associates, Washington, DC. http://www.usmayors.org/trees/treefinalreport2008.pdf.
55. Wolch, J.R., J. Byrne, J.P. Newell. 2014. Urban green space, public health, and environmental justice: The challenge of making cities "just green enough." Landscape and Urban Planning 125:234–244.
56. Zhang, Y., A. Hussain, J. Deng, and N. Leston. 2007. Public Attitudes Toward Urban Trees and Supporting Urban Tree Programs. Environment and Behavior 39(6):797-814.
57. Ziederman, A. 2006. Ruralizing the City: The Great Migration and Environmental Rehabilitation in Baltimore, Maryland. Identities 13(2):209-235.

PART III

MANAGING URBAN FORESTS

CHAPTER 6

A Protocol for Citizen Science Monitoring of Recently-Planted Urban Trees

JESSICA M. VOGT AND BURNELL C. FISCHER

6.1 INTRODUCTION

In the last decade, efforts are beginning to converge to monitor the survival, growth, and longevity of planted urban trees. In a comprehensive review of published single-tree inventory methodologies used in urban forestry (including aerial and satellite methods as well as traditional ground survey inventory methods), Nielsen et al. (2014) found that traditional "field survey," or on-the-ground, inventory methods constituted the vast majority of single-tree inventory studies (46 of 57 articles reviewed). Several recent large-scale, single-city tree-monitoring efforts have used field survey methods to measure the survival rates of urban trees. In the summer of 2006, the Parks and Recreation Department of New York City conducted a large-scale young street tree mortality study to examine the many factors in the city influencing the survival of over 14,000 newly

---

planted street trees (NYC Parks 2014). The site assessment tools used in this study included factors measuring the surrounding social and physical environment of each tree (NYC Parks et al. 2010). Other recent regional monitoring efforts include Sacramento, California, where Roman monitored the survival rates over 5 years of over 400 trees that were handed out as part of a utility company tree distribution program (Roman 2013); Milwaukee, Wisconsin, where, most recently, Koeser et al. (2013) use 25 years of monitoring data for a cohort of nearly 800 trees to determine the impacts of a variety of factors on tree survival rates; and New Haven, Connecticut, where Jack-Scott et al. (2013) evaluate the impact of community and other characteristics on survival rates for almost 1,400 trees planted between 1995 and 2007. To our knowledge, large-scale, multi-city planted tree monitoring studies do not seem to exist.

Standards for monitoring tree survival and growth over time are important for comparing the data obtained through different monitoring efforts across multiple locations and years (Leibowitz 2012; Roman et al. 2013; Nielsen et al. 2014). In 2011, the International Society of Arboriculture and The Morton Arboretum convened an international meeting on the subject of urban tree growth and longevity (Leibowitz 2012). This meeting organized four topic areas around descriptive studies of tree growth and longevity, plus three categories of factors influencing urban tree outcomes: tree production and sales, site design and tree selection, and tree and site management (Liebowitz 2012). The Urban Tree Growth and Longevity (UTGL) Working Group that emerged out of this meeting has undertaken to develop of a set of standards for monitoring the survival and growth of planted urban trees, as well as the factors that may influence survival and growth (UTGL Working Group 2014a). The Urban Tree Monitoring Protocol, as these standards are called, considers the factors of the tree, site, community, and management that may relate to tree survival and growth (UTGL Working Group 2014b).

We present in this paper the Planted Tree Re-Inventory Protocol for citizen science-based monitoring of recently-planted urban trees. Although we are members of the UTGL Working Group and the Urban Tree Monitoring Protocol committees, the protocol presented here was originally developed prior to the creation of the UTGL Working Group. Although our protocol and the in-progress UTGL monitoring protocol are informed by one another, our protocol is distinct in that it explicitly presents a data collection methodology for use by non-experts (i.e., citizen scientists) to measure trees in the urban landscape that have been planted relatively recently (trees in the establishment1 and semi-mature phase).[2]

This paper proceeds as follows: First, we reflect on tree planting organizations and their desire and capacity for monitoring. Then we define citizen science and review its use in urban forestry to date. Next, we discuss the measurement of urban tree outcomes (survival and growth) and summarize the literature on factors influencing tree success and urban forest outcomes. Finally, we present an overview of the main categories of variables included the protocol. The entire protocol is available on the Bloomington Urban Forestry Research Group website (http://www.indiana.edu/~cipec/research/bufrg_protocol) and as an appendix to this paper.

6.2 THE TREE-PLANTING ORGANIZATION CONTEXT

In 2010, our research group (the Bloomington Urban Forestry Research Group [BUFRG] at the Center for the Study of Institutions, Population and Environmental Change at Indiana University) was approached by the nonprofit urban greening organization, Keep Indianapolis Beautiful, Inc. (KIB), who was curious about the survival and growth of their planted trees. KIB works with neighborhoods and other groups to plant 1-2" (2-5 cm) caliper trees in the greater Indianapolis and Marion County, Indiana, area. They collect information about the location of each planted trees using global positioning system (GPS) units, and combine this with information obtained from the nursery about the species, planting packaging, and size (caliper, container size, etc.) of the trees they plant using a custom, self-designed Microsoft Access-based data management system. KIB lacked the resources to follow-up and monitor the survival, growth, and condition of these planted trees over the trees' early years (i.e., during the establishment and semi-mature phases before the trees reached their mature size). Their interest was twofold: First, KIB wanted to learn more about the survival and growth of trees they plant, and about the factors influencing the success of these trees. Second, and more importantly, KIB was looking for a way to expand capacity to monitor their planted trees into the future.

With KIB and other urban tree-planting organizations (including citizen groups, municipalities, etc.)[3] in mind, BUFRG embarked on the task of designing a method for reinventorying recently-planted urban trees that could be used by minimally-trained data collectors, ranging from high school students to casual adult volunteers. That our methods for inventorying planted trees be usable by non-expert individuals with minimal to no training in urban forestry or arboriculture—i.e., citizen scientists—was of key importance to our research group and to our main stakeholder, KIB. The resulting Planted Tree Re-Inventory Protocol enables citizen scientists to collect information about planted tree success (survival, growth and condition) as well as the factors that may influence tree success. Usability by citizen scientists makes our Protocol unique from existing urban forestry inventory protocols or standards.

6.3 CITIZEN SCIENCE

Citizen science, broadly defined, is the involvement of nonprofessional and amateur scientists— the average citizen—in scientific research efforts (Dickinson et al. 2012; Miller-Rushing et al. 2012; Shirk et al. 2012; Bonney et al. 2014). Citizen scientists can be paid interns, temporary workers or unpaid volunteers, and their efforts can augment data collection efforts undertaken by trained researchers, and thus expand the production of knowledge. Citizen science can involve a wide range of activities and various relationships between scientists and the general public. Miller-Rushing et al. (2012) describe three types of citizen science efforts, based on the level of public participation in the research process:[4] contributory (public contributes to data collection efforts only), collaborative (involving the public in data collection and also some parts of data analysis and results reporting), and co-created (public involved in all or most parts of the research process, from generating research questions to analyzing and reporting results).

True citizen science—like all science—involves a research question. Most projects in urban forestry are versions of Miller-Rushing et al.'s (2012) contributory citizen science that may or may not involve the processing and analysis of data to answer a true research question. These projects typically involve the public as members of urban forest inventory teams or in other monitoring efforts that might otherwise have been undertaken by urban forestry practitioners and certified arborists. Practitioners undertake inventories for a number of management purposes, including monitoring the success (survival and growth) of a group of trees over time, generating information about survival rates for use planning future tree planting efforts, providing information about the maintenance needs of a tree population, and more. All of these uses of inventory data center on the idea of adaptive management. Adaptive management occurs when the strategies used by resource managers are almost viewed as experiments or means of testing predictions about the relationships between management and a desired outcome (Holling 1996). Nonprofits or municipal forester managers that change the management strategies they use to plant or maintain trees based on the observed conditions of the urban forest as seen in urban tree inventory data are using adaptive management.

The use of volunteers to collect inventory data is not new in urban forestry. Tretheway et al. (1999) summarize the results of workshop on "Volunteer-Based Urban Forest Inventory and Monitoring Programs" convened by the U.S. Forest Service Pacific Southwest Research Station in 1999. Workshop participants identified three purposes for involving volunteers (i.e., citizen scientists) in urban forestry: to "provide a direct connection" between the community and the urban forest, to increase public awareness of the benefits and value of the urban forest, and to enhance support for urban forest "planning, management and stewardship" (Tretheway et al. 1999: p. 2). Cowett and Bassuk (2012) make the case for using university students at land grant colleges to conduct inventories; their "Student Weekend Arborist Teams" conducted more than 40 street tree inventories in small communities across New York State. Bancks (2014) discussed a University of Minnesota extension program that trains volunteers in communities of all sizes in urban forest rapid inventory methods, with the intent of assessing preparedness for emerald ash borer (see also http://mytreesource.com; University of Minnesota et al. 2014). Clarke (2009) describes the use of citizen science to track phenological trends in the urban forest as part of a larger citizen science program, Project BudBurst, managed by the U.S. Forest Service.

When research relies on citizen science for data collection, there can be concerns with the quality of the data collected. Several authors raise concerns about the accuracy of data collected by non-professionals (e.g., Dickenson et al. 2012; Roman et al. 2013). Bloniarz and Ryan (1996) evaluated the accuracy of inventory data collected by volunteers and found it to have similar levels of accuracy and consistency as data collected by certified arborists. In a more recent similar study, Bancks (2014) also found acceptable levels of accuracy for urban forest inventory data collected by volunteers. Future citizen science data collection efforts should continue to monitor the accuracy of data collected to ensure that it meets the quality required for good research.

Citizen science has the potential to substantially expand our ability to not only measure and monitor planted urban trees through time, but to also learn more about the factors influencing tree outcomes. Forty-two percent of practitioner-driven tree monitoring organizations surveyed by Roman et al. (2013) already make use of volunteers. And many tree-planting organizations already keep records with at least some information about the trees they plant (Roman et al. 2013). Rigorous citizen science tools that allow the public to record additional information about planted urban trees could help enhance both the quantity and quality of data on the urban forest available to tree planting organizations, tree managers, researchers, and decision makers. For instance, PhillyTreeMap (http://www.phillytreemap.org) is an urban tree mapping and monitoring project involving collaboration between multiple stakeholders in the Philadelphia area, including Azavea (a geographic information systems software and analysis company), Pennsylvania Horticultural Society (a tree-planting nonprofit organization), and the City of Philadelphia Parks and Recreation department, among other partners (Urban Forest Map et al. 2014). The PhillyTreeMap website and mobile applications allow individuals to enter information about a tree, including species, diameter, and height, and to view the amount of ecosystem services that tree and other trees in the database provide.

The implementation of similar tree-monitoring projects across multiple cities and regions and the integration of data collection methods for more information about each tree would enhance the appeal of volunteer-generated datasets to researchers interested in answering explicit research questions. More direct connections and collaborations between practitioner-driven inventory efforts and researchers would truly launch urban forestry into the land of citizen science. New technologies for monitoring may even allow urban tree monitoring to eventually rival "big data" citizen science projects like Galaxy Zoo (http://www.galaxyzoo.org; Zooniverse 2014) and the Christmas Bird Count (http://birds.audubon.org/christmas-bird-count; National Audubon Society 2014).

6.4 MEASURING URBAN TREE OUTCOMES

Whether as trained experts or citizen scientists, when we measure urban forest outcomes at the level of the individual tree, there are two different general approaches: place-based inventories and cohort studies. Place-based inventories aim to capture information about a particular type of trees in a given area (e.g., street trees on a major street, public trees in a single neighborhood, all trees on a particular piece of property). Inventories are the more common approach to measuring the urban forest, and street tree inventories in particular have been the norm for capturing the information necessary to calculate the benefits of the urban forest. Cohort studies take a different approach: instead of measuring a particular type of trees in a single area, these studies monitor a cohort—or group of trees planted at the same time—through multiple years or at multiple future points in time. Cohort studies may follow all the trees planted as part of a neighborhood tree-planting project, annual tree-planting program by a municipality or nonprofit, tree distribution program, or other event where multiple trees were planted at the same time, and there is interest in tracking the outcomes of the planted trees over time. For cohort studies, we usually know the actual date, season, or year of planting for each tree, whereas for inventories the date of planting is likely unknown.

Whether tracking a single cohort of trees planted at the same time or inventorying all the street trees in an entire city, we are measuring features of each individual tree in the inventory. At the level of the individual tree, urban forest outcomes can be operationalized several ways: we could measure tree health, vigor, or condition; the amount or value of benefits produced by a tree; tree size or growth rate; or, most simply, whether or not a tree lives or dies. Here, we discuss tree survival (or conversely, mortality) and tree growth rates, as two of the more common tree-level outcomes.

6.4.1 URBAN TREE SURVIVAL (AND MORTALITY)

A common urban forest axiom is that the expected life of a street tree is only 7 or 10 years, but Roman and Scatena (2011) acknowledge that it's unclear where this life expectancy estimate comes from, and provide a more empirical estimate of 19 to 28 years. There are a number of types of mortality for trees in urban areas. Clark and Matheny (1991) identify three primary reasons that trees die in urban areas: structural failure, environmental degradation, and parasitic attack. Different types of mortality may be more closely linked to certain stages in a tree's lifecycle, and so another typology of mortality might be establishment-related mortality, damage-related mortality, and age-related mortality. Establishment-related mortality is connected to the tree's failure to establish in the landscape after transplanting, either due to inadequate care (i.e., not watered after planting), poor tree stock, or improper site selection (not the "right tree" in the "right place"). Damage-related mortality is the death of a tree directly due to damage by humans, either during construction of roads, buildings, or other urban infrastructure that results in removal of the tree during or after the construction, or other damage (due to an automobile, lawnmower, etc.) that necessitates the tree's removal. Age-related mortality is the typical cause of death for non-urban trees; age-related death results from the natural senescing process undergone by trees, through which first small branches and then large branches and then the whole tree stop producing new growth or green leaves every season. Age-related mortality is closely connected to mortality caused by pests or diseases, which are more likely to affect declining or already dying trees.

When calculating a mortality rate for a group of planted trees, unless the cause of tree mortality or failure was recorded for each tree (i.e., as in the case of trees in the International Tree Failure Database; ITFD 2014), most of the time we cannot distinguish the types of mortality from one another. Especially in cases where the tree has been removed, the only thing monitors can know is that where there was once a tree, there is no longer a tree. For this reason, defining "mortality rates" for a cohort of planted trees or for an inventory becomes rather muddled. We cannot know, for instance, what portion of the calculated mortality rate is due to the planting of poor nursery stock relative to what is due to activities undertaken (or not) post-planting in the name of tree care. Long-term data on the same trees at multiple points in time generated through citizen science-based monitoring efforts can help fill this gap in our knowledge.

6.4.2 GROWTH

Urban tree growth is another measurable urban forest outcome. Large, mature trees provide many more benefits than small or immature trees; thus, the faster a tree grows, the sooner it will yield a return on investment (Nowak et al. 1990). Growth rates are measured a number of different ways in the urban forestry literature, including change in tree height (e.g., Stoffberg et al. 2008; Jutras et al. 2009), amount of new shoot growth at the ends of branches (e.g., Solfjeld and Hansen 2004), change in diameter at breast height (e.g., Nowak, McBride & Beatty 1990; Jack-Scott et al. 2013), change in caliper (diameter at 6 in [15 cm] above the first lateral root; e.g., Struve et al. 2000), and the width of annual growth rings as obtained from tree cores (e.g., Iakovoglou et al. 2001). Peper and McPherson (2003) evaluated several methods for measuring leaf area of urban trees that could be used to measure or model canopy growth and change. There are relatively few studies of urban tree growth—particularly longitudinal studies (Liebowitz 2012). And although tree growth has been examined in nursery and experimental settings, few researchers have examined urban tree growth in situ in actual cities.

6.5 FACTORS THAT INFLUENCE URBAN TREE OUTCOMES

Tree survival (mortality) and growth is influenced by a large number of factors. Several existing organizing frameworks can be helpful in identifying categories of variables that might influence urban tree outcomes. The social-ecological system (SES) framework developed in rural natural resource management settings states that the characteristics of the resource itself (for example, a forest), the resource system (the trees), the resource users or actors (timber harvesters), and their governance system (rules about when and how to cut trees) influence outcomes observed in coupled human-natural systems (e.g., Ostrom 2009; Ostrom and Cox 2010). In urban forestry, the Clark et al. (1997) "Model of Urban Forest Sustainability" states that sustainable urban forest outcomes are predicated on "a healthy tree and forest resource, community-wide support and a comprehensive management approach" (Clark et al. 1997, 17). Tree biologists and plant physiologists also delineate categories of variables that influence plant growth. In *Growth Control in Woody Plants*, Kozlowski and Pallardy (1997) review the numerous factors influencing tree and shrub growth. These authors outline categories of physiological factors, environmental factors, and "cultural practices," and describe how each category influences the reproductive (production of flowers and pollen, fertilization and eventually fruiting) and vegetative (root and shoot) growth of woody plants (Kozlowski and Pallardy 1997).

We combine these ideas into an interdisciplinary[5] social-ecological systems perspective of urban forest outcomes (Table 1, Figure 1). Adapted from SES theory (Ostrom 2009) and the Clark et al. (1997) model, and informed by tree physiology research (Kozlowski and Pallardy 1997), Table 1 presents urban forest outcomes as the product of interactions between the components of the urban forest social-ecological system. Thus, urban forest outcomes— including tree survival, growth, condition, etc.—are influenced by the interactions between the characteristics of the tree itself, the biophysical environment, the community, and the institutions and management strategies (Figure 1).[6]

The following section describes the current state of knowledge for each of the four main categories of variables that influence tree outcomes. This abbreviated literature review uses the three key sources from Table 1 (Clark et al. 1997; Kozlowski & Pallardy 1997; Ostrom 2009) as well as other relevant literature from the fields of urban forestry/arboriculture, urban ecology, natural resource management, coupled human-natural systems, and more.

Table 1. The urban forests as social-ecological systems perspective draws on several organizing frameworks, including the Model of Urban Forest Sustainability (Clark et al. 1997), the Social Ecological Systems (SES) Framework (first developed by Ostrom [2009], but see also Ostrom & Cox [2010]), and Kozlowski and Pallardy's (1997) Growth Control in Woody Plants. *"Institutions" refers to the rules and shared strategies (per Ostrom 2005) used by people to manage and maintain trees as well as the surrounding biophysical environment in the urban forest. [Modified from http://www.indiana.edu/~cipec/research/bufrg_about.php).

Social-Ecological Systems Framework	Model of Urban Forest Sustain-ability	Growth Control of Woody Plants	Urban Forests as Social-Ecological Systems
Resource Units	Vegetative Resource	Physiology	Trees
Resource System		Environment	Biophysical Environment
Governance System	Resource Management	Cultural Practices	Institutions & Management
Resource Users or Actors	Community Framework	--	Community

6.5.1 CHARACTERISTICS OF THE TREE

The characteristics of the tree itself obviously impact its survival and growth. Clark et al. (1997) use vegetation resource to refer to the trees in the urban forest, listing canopy cover, age distribution, species mix, and native vegetation as the key features of the urban forest that influence its sustainability. Here, we focus on the characteristics of an individual tree—including physiology—that influences its success. Kozlowski and Pallardy (1997) discuss the following key physiological processes as they relate to tree growth: production of carbohydrates via photosynthesis, mineral uptake and use, internal water relations and evapotranspiration, and hormone regulation. Clark and Matheny (1991) note that a tree's growth rate depends significantly on the availability of resources (carbohydrates, minerals, water, etc.) and that when resources become limiting growth is reduced. Because these physiological processes that manage resources

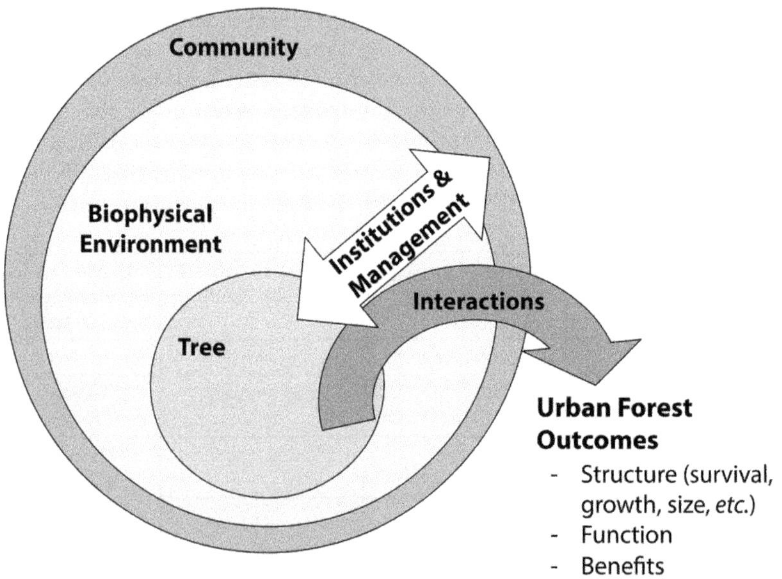

FIGURE 1: The urban forests social-ecological systems perspective emphasizes that the community interacts with trees and the biophysical environment through institutions and management to produce outcomes in the urban forest.

are clearly connected to tree genetics, it should come as no surprise that different species exhibit different survival and growth rates (e.g., Iakovoglou et al. 2001; Grabosky and Gilman, 2004). For transplanted trees, the physiological processes that impact tree establishment, survival, and growth in the landscape are affected by characteristics of the tree at the time of transplanting. The size of the tree at planting has been linked to subsequent survival and growth (Lambert et al. 2010). Nursery production method and the type of plant packaging can also impact transplanted tree success (Gilman and Beeson 1996; Buckstrup and Bassuk 2000). Trees planted too deeply or with excessive mulch covering the rootball exhibit higher mortality rates than trees planted at the proper depth (Gilman and Grabosky 2004). Tree condition and health are also linked to tree success. Lower tree condition ratings are associated with decreased odds of tree survival (Koeser et al. 2013) and lower growth rates (Berrang et al. 1985).

6.5.2 BIOPHYSICAL ENVIRONMENT

Factors in the surrounding biophysical environment also influence tree outcomes. Environmental factors include variables that might be studied by a plant ecologist, such as light availability and intensity, water relations (including drought and flood conditions), temperature, soil nutrient content and physical structure (e.g., compaction), pollution, and other abiotic (e.g., wind, fire) and biotic (pests and diseases) factors (Kozlowski and Pallardy 1997). The biophysical environment may have a particularly strong effect on urban tree success, and street trees in particular experience stressful growing conditions. The most influential environmental factors are significantly different for trees in urban areas compared to rural, more natural growing environments. Urbanization increases impervious surfaces, buildings, and other built or grey infrastructure, resulting in radical changes in the water, temperature, and other abiotic conditions across the urbanized landscape (Arnold and Gibbons 1996; US EPA 2008). Water stress is commonly cited as a limiting factor for urban tree growth (Kramer 1987; Krizek and Dubik 1987; Graves et al. 1991), particularly in arid regions (Costello 2013; Symes and Connellan 2013). High air temperatures can disrupt tree phenology and reproductive growth, and higher soil temperatures can change seasonal root growth patterns (Kozlowski and Pallardy 1997). Because water availability, temperature, and other characteristics of the biophysical environment vary throughout the year for most locales, the season of planting may also impact tree outcomes (Anella et al. 2008; Solfjeld and Hansen 2004). Additionally, several authors have found that smaller available rooting volume leads to constrained root, trunk, and shoot growth (Krizek and Dubik 1987; Grabosky and Gilman 2004; Day et al. 2010). Competition with other trees for space, nutrients, light, water and more can also limit tree growth and survival (Nowak et al. 1990; Rhoades and Stipes 1999; Iakovoglou et al. 2001). Compounding space constraints are the generally poor soil conditions in urban areas (Scharenbroch et al. 2005; Smith et al. 2001).

The urban forest axiom right tree, right time, right place is often on the minds of tree planters, and even sometimes a piece of urban forest policies, plans, or ordinances. Several efforts are currently underway to develop a more empirical foundation to the linkages between tree outcomes and site and soil characteristics, including work led by Bryant Scharenbroch at the The Morton Arboretum (MASS Laboratory 2014). Our protocol includes measurement of variables that are proxies or indicators for available growing space above and below ground and the quality of the site.

6.5.3 INSTITUTIONS AND MANAGEMENT

Tree success is also impacted by the institutions—i.e., management strategies and maintenance practices—that arborists, urban foresters and other members of the community use to care for urban trees. Kozlowski and Pallardy (1997) refer to these activities as cultural practices that influence tree growth, and their list includes typical tree maintenance activities such as pruning and watering, use of fertilizers, growth regulators, or other chemicals, spacing of trees (both initial arrangement of planted trees and thinning of existing forest stands), and, even protection from freezing. The Clark et al. (1997) resource management component includes mostly variables representing administrative or organizational features of city government as these might relate to adequacy of resources for urban tree management: city-wide management plan, funding, staffing, assessment tools, protection of existing trees, species and site selection, standards for tree care, citizen safety, and recycling. The SES framework (e.g., Ostrom 2009) uses the term institutions to refer to the formal and informal rules and shared strategies that structure the interactions among individuals and groups of people and between people and their environment (Ostrom 2005).

Much of the research on institutions emerges from studies on common pool resource (CPR) management conducted in the disciplines of political science, economics, and anthropology (e.g., Ostrom 1990, 2005). Theory on CPR management states that several principles are likely to be linked to persistent or sustainable systems, including effective monitoring, appropriate sanctioning of rule-breakers, rules allowing individuals impacted by the resource and rules to change those rules, and strategies for effective conflict resolution (Cox et al. 2010). Institutions as rules have only been cursorily examined in urban ecosystems, and not at all for urban forest outcomes (Mincey et al. 2012). Larson et al. (2008) describe rules of homeowners' associations that limited the appearance and management strategies used for residential vegetation, including pest and water management methods and species composition. Mincey and Vogt (2014) find that watering strategy used by the neighborhood impacts tree survival rates.

Tree maintenance strategies can be characterized by the type of maintenance (e.g., pruning, watering), intensity (how much maintenance is performed, i.e., training pruning, 15 gallons of water), frequency (how often the activity is performed, e.g., annually, once per every week it does not rain), duration (how long the activity is performed, e.g., for the first 5 years after transplanting), and extent (which trees or what part of each tree is maintained, e.g., pruning up lower branches, watering all trees in the State St. right-of-way) (Vogt, Hauer, and Fischer in review). Maintenance type, intensity, frequency, duration, and extent all influence tree and urban forest outcomes; the impact of watering (Gilman 2001, 2004), pruning (Whitcomb 1979; Miller and Sylvester 1981; Evans and Klett 1985), and mulching (Gilman and Grabosky 2004) varies depending on the particulars of the maintenance strategy.

Maintenance strategies or institutions or rules about tree care may not always be visible on the tree itself or in the area nearby. Our Protocol includes a few key maintenance practices— pruning, mulching, staking—of which evidence can be seen on the tree itself.

6.5.4 COMMUNITY

Because urban trees are surrounded by people, the characteristics of the community of people living in and around the urban forest influence tree outcomes. For instance, Boyce (2010) observed that the designation of volunteer tree stewards in the community dramatically reduced urban tree mortality rates. The components of community framework included in the Clark et al (1997) model are public agency cooperation, involvement of large private and institutional landholders, green industry cooperation, neighborhood action, citizen-government-business interaction, general awareness of trees as a community resource, and regional cooperation.[7]

Most of the empirical evidence for the influence of community characteristics on environmental outcomes emerges from the research that informed development of the SES framework. Because of its emphasis on rural natural resource management, the SES framework uses the terms "resource users" or "actors" to describe the community of people that manage and use a resource (Ostrom 2009; Ostrom and Cox 2010; Epstein et al. 2013). Features of the community that impact resource management outcomes according to the SES framework include community size (population or number of people involved), history using or managing the system (i.e., experience), demographic or socioeconomic characteristics, individual knowledge (of the resource system), norms (individual perceptions of socially-acceptable practices), and the location of the community (Ostrom 2009; Ostrom and Cox 2010; Epstein et al. 2013).

Some of the resource user or actor characteristics listed above have been examined for urban forest social-ecological systems. Iakovoglou et al. (2002) find no significant difference in growth rates between different-sized communities. Jack-Scott et al. (2013) found that a greater number of participants in tree planting events during a year is associated with higher survival and growth rates. Land use type is a factor partially indicative of the features of the biophysical environment but perhaps more closely captures community characteristics. Several authors have found an effect from adjacent or surrounding land use type on tree success (Nowak et al. 1990; Lu et al. 2011). A few studies have found that demographic characteristics (i.e., variables from the U.S. Census) are related to tree outcomes (Nowak et al. 1990; Grove et al. 2006). Lastly, studies from the field of urban ecology have observed that norms or individual motivations impact landscape outcomes (Austin 2002; Grove et al. 2006; Nassauer et al. 2009).

Like institutions, characteristics of the community of people are difficult to observe during on-the-ground inventory. Our Protocol adapts several of the stewardship factors collected by the New York Young Street Tree Mortality Study (NYC Parks et al. 2010) as indicators of a care ethic in the community surrounding the tree.

6.5.5 INTERACTIONS AND ENDOGENEITY

Complex coupled human-natural systems are inherently filled with endogeneity, or simultaneous interactions between variables that complicate and sometimes obscures our understandings of the causal impact of variables on observed outcomes (Liu et al. 2007; Schlüter et al. 2014). The urban forest social-ecological system is no exception: interactions within and between tree, biophysical environment, community, and institutional factors can influence urban forest outcomes as much as the influence of a single factor. For instance, proper, proactive maintenance strategies may actually mitigate the impact of sub-optimal growing conditions. Additionally, alignment between rules, the characteristics of the community and local conditions has been demonstrated to impact common-pool resource outcomes (Cox et al. 2010). And characteristics of the community such as individual preferences and knowledge may impact choice of management strategies. A study of residential yards in Minnesota found that homeowners' application of water, fertilizers, and weed killers, as well as other yard management techniques was strongly influenced by resident knowledge and perception of the yard as a relatively closed system (Dahmus and Nelson 2013). Additionally, Vogt et al. (in review) observed an interaction between watering strategy and planting season.

6.6 THE PLANTED TREE RE-INVENTORY PROTOCOL

In light of these four main categories of variables that influence urban forest outcomes, we present here the Planted Tree Re-Inventory Protocol (see the Appendix of this paper for Version 1.1; refined from an earlier version of the protocol: Vogt et al. 2013). The protocol describes standardized methods that can be used by non-professional inventory personnel to gather data necessary to evaluate the survival and growth of recently-planted[8] urban trees, as well as the many factors influencing survival and growth.

Selection of variables to include in the protocol was informed by the literature review summarized above as well as existing urban tree inventory methods, including the i-Tree Eco field methods (i-Tree version 4.0 of the user's manual was consulted for this work), the *Standards for Urban Forestry Data Collection* (IUFRO et al. 2010), and the methods of New York City's Young Street Tree Mortality Study (NYC Parks et al. 2010; results summarized by Lu et al. 2010). Individual variables and values of each variable were debated by members of the Bloomington Urban Forestry Research Group (BUFRG) over the course of a 6-month period following the review of literature and inventory methods. Table 2 lists each of the variables in the final protocol and, if applicable, the original source for their methods. We adapted and modified variables from other inventory methods to make sure that each variable could be successfully assessed by minimally-trained data collectors. To this end, many variables in the protocol require only simple, qualitative, visual assessments of the tree and its environment, and not precise measurements. For instance, a simple presence or absence assessment method, where the data collector only has to determine whether or not a particular feature is present or absent on the tree or nearby surrounding environment, is used for many variables. Variables that do ask for more precise quantification (e.g., measurements of diameter, height, or distance) require use of only two or three simple tools: a diameter tape and a digital range finder (hypsometer) or clinometer and measuring tape.

The protocol was tested by several different parties (Table 3). A preliminary list of variables was tested by members of BUFRG in the summer of 2011. Since the final users of the protocol were to be minimally-trained, non-professional data collectors, high school members of KIB's Youth Tree Team (YTT) tested the protocol during the summer of 2012; YTT used a version of the protocol adapted for use on ESRI's ArcGIS iPhone mobile application to collect data for more than 700 recently-planted street trees. YTT data collection team members were trained in data collection methods during two 6-hour training days, and overseen by a collegeaged YTT Leader who had participated in approximately 15 additional hours of data collection activities with members of BUFRG during Protocol development. The YTT training procedures described above are similar to those used in studies that have found high accuracy for volunteercollected data (Bloniarz and Ryan 1996; Bancks 2014). The protocol was also tested on slightly more mature trees planted between

2000 and 2011 on City of Bloomington right-of-ways; IU master's students collected data on over 1,000 street trees using paper-and-pencil in the summer of 2012.

In addition to collection and evaluation of tree data using the protocol, testing also consisted of written daily field notes taken by YTT members (Vogt et al. 2012) as well as extensive informal discussion between members of the YTT team engaging in data collection and the researchers. For instance, the original protocol called for collecting presence or absence information on several different leaf conditions (evidence of insects, rust, chlorosis, and other leaf condition notes); however, based on written field notes from YTT members, we reduced leaf condition variables to just one: chlorosis. We also clarified that to be considered "present," chlorosis must be evident on at least 25% of the leaf surface area of the tree, and provided pictures and sketches of chlorosis to help with identification and estimation. Written field notes feedback also encouraged us to clarify instructions provided for locating each tree. Additionally, at the end of the data collection season YTT members narrated their thinking while collecting data into an audio recorder. This recording was used to verify that data collection methods had not changed between the beginning and end of the summer, and slight modifications were made to variable descriptions and instructions in the protocol based on decisions and strategies that data collectors were using in the field. (For example, narration revealed that data collectors were marking "incorrect mulching" for trees with very old, degraded mulch, where only few bark chips were still visible. The definitions of correct, incorrect, and no mulching in the protocol were updated to clarify that this case would actually better be classified as "no mulch," given the biophysical implications of capturing information about correct versus incorrect mulching.)

Version 1.1 is presented here. In the remainder of this paper, we briefly describe the variables included in the protocol. The entire protocol (in PDF form) is available as a supplementary online appendix to this article, as well as on the BUFRG website (http://www.indiana.edu/~cipec/research/bufrg_protocol.php) in both greyscale and color versions, along with a quick reference guide for the field and customizable and printable data collection sheets.

6.6.1 TREE CHARACTERISTICS

Biophysical variables (tree characteristics and local environmental variables) compose the majority of the variables in most tree inventory protocols, including this one, for a couple reasons: first, factors about the tree and immediate surroundings are most easily observed by data collectors. Second, most tree inventory methods used by urban foresters and arborists are informed by forest mensuration methods used in traditional forestry. Third, as noted above, most research on urban tree survival and growth has emerged from the fields of horticulture and arboriculture, and these fields are strongest in their assessment of the impact of tree and environmental factors on growth.

6.6.1.1 IDENTIFYING INFORMATION

The most critical information collected in any inventory protocol is basic identifying information about the tree. This includes a tree identification number, some sort of location information, and species. An identification number is a unique value for each tree in the inventory, useful for tracking the same tree over time through multiple inventory years. Location information should include enough information so that the physical location of the tree in space can be found. Location may be an address number and street name of the property adjacent to the tree, geographic coordinates (I.e., GPS latitude and longitude), distance and direction of the tree from the nearest street intersection, or any other way to precisely locate the tree. Species is the biological name for the type of tree that was planted. Species can be detailed, and include the cultivar or variety (e.g., autumn blaze maple, *Acer x freemanii* 'Jeffersred'), or could be limited to just the genus (e.g., *Acer* spp.) of the tree planted, depending on the level of detail desired for the inventory and the tree identification skills of data collectors.

6.6.1.2 SIZE

In order to measure growth of trees over time, we need information about trees' size. Size information included in the protocol is diameter at breast

Table 2. Original sources for variables included in the Planted Tree Re-Inventory Protocol. Complete citations in Literature Cited.

VARIABLE NAME		ADAPTED/MODIFIED FROM (if applicable)
Tree characteristics		
Identifying information		
V1	Tree ID	
V2	Location	
V3		
Size		
V4	DBH	IUFRO et al. 2010: p. 2-3
V5	Caliper	
V6	Total height	IUFRO et al. 2010: p. 3
V7	Height to crown	IUFRO et al. 2010: p. 3-4
Canopy		
V8	Crown dieback	IUFRO et al. 2010: p. 8
V9	Crown exposure	IUFRO et al. 2010: p. 4-5
V10	Chlorosis	
Trunk		
V11	Root flare	IUFRO et al. 2010: p. 23
V12	Lower trunk damage	
Overall condition		
V13	Other damage	
V14	Overall tree condition	Fischer et al. 2007: appendix
Local environment		
Near tree		
V15	Utility interference	IUFRO et al. 2010: p. 9
V16	Building interference	IUFRO et al. 2010: p. 9
V17	Fences interference	IUFRO et al. 2010: p. 9
V18	Sign interference	IUFRO et al. 2010: p. 9
V19	Lighting interference	IUFRO et al. 2010: p. 9
V20	Pedestrian traffic interference	IUFRO et al. 2010: p. 9
V21	Road traffic interference	IUFRO et al. 2010: p. 9
V22	Ground cover at base	IUFRO et al. 2010: p. 14
V23	Ground cover under canopy	IUFRO et al. 2010: p. 14

Table 2. Continued.

Planting area		
V24	Planting area type	
V25	Planting area relative to road	
V26	Planting area width	IUFRO et al. 2010: p. 15-16
V27	Planting area length	
V28	Curb presence	NYC Parks and Recreation et al. 2010: p. 20
Proximity to other things		
V29	Number of trees in 10-m radius	Iakovoglou et al. 2001: p. 75
V30	Number of trees in 20-m radius	Iakovoglou et al. 2001: p. 75
V31	Number of trees in same planting area	
V32	Distance to road	IUFRO et al. 2010: p. 16
V33	Distance to building	IUFRO et al. 2010: p. 9
Management		
Maintenance		
V34	Pruning	NYC Parks and Recreation et al. 2010: p. 22
V35	Mulching	
V36	Staking	
Community		
Evidence of care		
V37	Water bag	NYC Parks and Recreation et al. 2010: p. 22
V38	Bench	NYC Parks and Recreation et al. 2010: p. 22
V39	Bird feeder	NYC Parks and Recreation et al. 2010: p. 22
V40	Yard art	NYC Parks and Recreation et al. 2010: p. 22
V41	Trash/debris	NYC Parks and Recreation et al. 2010: p. 22

height, caliper, total height, and height to crown. Diameter at breast height (DBH, or diameter measured at 4.5 ft or 1.3 m off the ground) is one of the most commonly used metrics of size for trees in rural or urban areas. The change in DBH over time is one way to calculate tree growth, and DBH can also be used to calculate the total benefits provided by the tree (e.g., carbon storage). Caliper, or tree diameter 6 inches (15 cm) from the first lateral root, can also be used to calculate tree growth. This is a particularly convenient measure for recently-planted trees, because trees are often sold from the nursery by caliper size; comparing current caliper with that from

Table 3. Protocol testing sites, trees, and data collectors. **Living trees* indicates that only trees remaining at the time of re-inventory were assessed using the Protocol. *Planted trees* indicates that all trees planted were assessed (i.e., for trees removed since planting, the Overall tree condition was assessed as "Missing" and only select biophysical environment variables were collected).

Site	Number of trees*	Tree planting years	Trees planted by	Data collectors	Data collection dates
Indianapolis	120 living trees	2006-2007	Volunteers of Keep Indianapolis Beautiful	IU BUFRG researchers	June-Sept 2011
Bloomington	1,097 planted trees	2000-2011	City of Bloomington Parks and Recreation Division of Urban Forestry	IU Master's of Science in Environ-mental Science students	May 2012
Indianapolis	714 planted trees	2006-2009	Volunteers of Keep Indianapolis Beautiful	High-school aged Keep Indianapolis Beautiful's Youth Tree Team (YTT) members led by a collegeage YTT leader	June-July 2012

the tree at the time of planting is another means of calculating tree growth. Total tree height and height to crown provide a metric of above ground size, and can be combined to provide a simple proxy for crown or canopy volume and potential for photosynthesis and growth.

6.6.1.3 CANOPY

Tree health and condition includes information about the canopy, trunk, and entire tree. Information about the condition of the canopy (or leafy top of the tree, also called the crown) is important for assessing the health of the tree. Canopy information included in the protocol is crown dieback rating, crown exposure rating, and presence of chlorosis. Crown dieback and exposure are qualitative visual assessments, recorded on simple point rating scales, using methods modified from the Urban Forestry Data Standards (IUFRO et al. 2010). Crown dieback is a qualitative assessment of the percent of dead branches in the canopy relative to the total living crown, assessed on a 0-6 scale. Crown exposure is a rating of how much of the tree's canopy is exposed to sunlight, based on how many sides of the canopy are shaded by buildings or other trees, assessed on a 0-5 scale. Chlorosis is a presence or absence metric, where "presence" implies that leaf chlorosis is evident on at least 25% of the leaf surface area of the entire tree.

6.6.1.4 TRUNK

Trunk condition metrics are equally as important as canopy condition in assessing overall health of the tree. Trunk condition is related to the health of its vascular tissue and the ability of a tree to successfully transfer nutrients and water between the root system and canopy. Trunk information included in the protocol is presence of a root flare and presence of lower trunk damage. A root flare, or gradual taper of the trunk of a tree as it enters the ground, may be indicative of how deeply the tree was planted.[9] The roots of trees planted too deeply may lack sufficient access to oxygen, may be more at risk of water stress (e.g., Gilman 2004) or may be prone to root girdling of the tree. Trees exhibiting lower trunk damage—such as that caused by a lawn mower, weed-whacker, or even animals—may be at greater risk of infection by fungus or disease. Repeated damage over time and on all sides of the lower trunk, such as from a lawn mower, may even girdle the tree, severing the vascular tissue and preventing water and nutrient transfer.

6.6.1.5 TREE CONDITION

Presence of any other damage and determining an overall tree condition rating are the final assessments of tree-level variables, made after both canopy and trunk condition as well as all other aspects of the individual tree have been examined. Other damage to the tree that may impact its health, condition, survival or growth include: broken branches, branches stripped of leaves or bark, damage to the upper trunk of the tree, a wire or other item choking or girdling the tree, etc. Overall tree condition takes into account the condition of the trunk and canopy. A deciduous tree in good health and condition exhibits a full canopy of dark green leaves that are not undersized for the current season, and a growth form appropriate for its species, without dead branches or excessive water sprouts growing out of the base or main trunk of the tree. Conifers in good health have full boughs with dark green needles. Tree condition ratings should consider a tree from all angles and from top to bottom. The protocol condition ratings range from good to dead and include categories for stumps, sprouts, or absent trees.

6.6.2 LOCAL ENVIRONMENT

6.6.2.1 NEAR TREE ENVIRONMENT

In the local environment immediately around the tree, we can assess the quality and quantity of growing space by assessing interference with infrastructure (utility, building, fences, sign, lighting, pedestrian traffic, and road traffic) and type of ground cover (at the base of the tree, and under the canopy). Interference with infrastructure is assessed according to whether or not the tree is in conflict with aboveground utility wires or poles, buildings, fences, signs, or lighting at the time of re-inventory. Interference with traffic refers to the presence of branches more than ½ inch (1 cm) in diameter at or below 8 ft (2.4 m) above a pedestrian walkway or sidewalk for pedestrian traffic, or, for road traffic interference, at or below 14 ft (4.3 m) above an active lane of traffic (i.e., not a parking lane). Trees that are located in close enough proximity to infrastructure so as to conflict with it may compete with this infrastructure for aboveground growing space, or may require more frequent pruning to limit conflicts between branches and the built environment. The type of ground cover around the tree is a qualitative assessment of the type of cover (e.g., bare soil, mulch, grass, etc.) at the base of as well as under the canopy of the tree. Ground cover reflects the surface conditions of the belowground growing environment, including potential competition with other plants for water and nutrients, the permeability of the area to infiltration of water, or even the likelihood of surface soil disturbing activities (such as digging in an annual flowerbed).

6.6.2.2 PLANTING AREA CHARACTERISTICS

The quality and quantity of growing space is also related to the planting area type, its position relative to the road, its length and width, and the presence of a curb at the edge of the planting area. Planting area type refers to the type of physical space in which the tree as planted; types of planting areas include a tree lawn, median, shoulder, tree grate, tree pit, bumpout, front yard, side yard, or other open area. Sketches of each type of planting area are provided in the protocol. The size of the planting area as measured by its surface area (length and width) is a proxy for available rooting space below ground. In addition to the type and size, the position of the planting area relative to the road (i.e., above, even, or below the surface of the road) as well as whether or not the planting area has a curb may impact the quantity and quality of any runoff into the tree planting area.

6.6.2.3 PROXIMITY TO OTHER THINGS

Other living and nonliving things in the larger growing area of the tree can also impact tree success. The protocol considers the number of trees in a 10-meter (33-ft) radius, a 20-meter (67-ft) radius, and the same planting area, as well as the distance to the nearest road and building. The number of other trees near the sample tree influence the amount of competition a tree experiences, both above and below ground, for light, nutrients, water, and growing space. The distance to the nearest road can tell us about potential exposure to factors that may influence a tree's health, condition or growing potential, including the potential for automobile injury or road spray contaminated by fuels, salts and other particles. The distance to the nearest building can tell us about the potential exposure to radiant building or for shading by the building.

6.6.3 MANAGEMENT VARIABLES

Most management and maintenance cannot be captured using on-the-ground tree inventory methods, but might be better captured through surveys or interviews of the individuals or groups responsible for the trees. However, some maintenance is visible when looking at the tree during an on-the-ground inventory. The protocol includes variables that consider evidence of pruning, mulching, and staking on the tree, as well as whether the maintenance activity appears to have been performed correctly or incorrectly. For instance, correct pruning cuts should be a smooth, flat cut, made just outside the branch collar for a branch off the main trunk of the tree, or just after the branching for secondary branches in the crown. The protocol includes sketches with examples of correct and incorrect pruning and mulching, and complete text descriptions for correct and incorrect pruning, mulching, and staking.

6.6.4 COMMUNITY VARIABLES

The last suite of variables included in the protocol considers the surrounding community as it is manifested in evidence of care around the tree. The protocol includes four indicators of positive norms of care—presence or absence of a water bag, bench, bird feeder, or yard art (adapted from the list considered by the New York City Young Street Tree Mortality study [NYC Parks et al. 2010])—and one indicator of a lack of care—presence of trash or debris.

6.7 CONCLUSION

Data collected via the protocol has many uses, depending on the end user. Tree planting organizations might use the data to help plan the locations and management of future tree planting efforts. Municipal urban foresters might use data on cohort survival rates to help determine an annual budget for planting new trees. Researchers might use data to better understand the myriad factors that influence urban tree outcomes and to create better models of tree growth and survival over time and to improve estimates of the benefits of the urban forest.

As urban areas continue to develop and redevelop, to expand and infill, the number of non-planted (i.e., remnant) trees in cities will continue to decrease, as relatively natural areas are replaced by designed landscapes of buildings, roads, planted trees, and other infrastructure (both green and grey). While cities and developers often maintain complete and detailed plans of buildings and roads, detailed records of planted trees rarely exist. However, trees are an integral part of urban infrastructure. In order to ensure they continue providing benefits to urban residents, we should keep track of the location, survival and growth of the trees we plant so that they can be efficiently managed and maintained throughout their lifetimes, and then removed and replaced after they die. With better data about planted urban trees, we can more efficiently allocate limited resources for managing and maintaining the urban forest.

The protocol methods presented in this paper can serve as a beginning of a conversation between researchers, urban forestry practitioners, and the public about the measurement of the factors that influence the success of recently-planted urban trees. The protocol will continue to be used and tested by various groups, and accuracy assessments of data collected by citizen scientists should be conducted. We expect to continue to publish new and updated versions of the protocol on the BUFRG website.

FOOTNOTES

1. The establishment phase is typically, 2 or 3 years for trees 3-5 cm (1-2") in caliper at planting.
2. Therefore, we do not include metrics commonly included in urban forest inventory methods, such as maintenance requirement variables or hazard/risk assessment methods, that may be both difficult for the non-expert to assess as well as not applicable to most immature trees.
3. KIB is not alone in their interest in tools for monitoring planted trees. In a survey of 32 practitioner organizations already engaged in monitoring efforts, Roman et al. (2013) observed a desire for simple protocols over those that are "complicated and academic" (p. 296). In the same survey, practitioners cited challenges associated with monitoring, including a lack of staff time and dedicated funding, finding and using technology resources, and developing or choosing appropriate protocols (Roman et al. 2013).
4. Shirk et al. (2012) define similar types of citizen science, and their classification also includes *contractual* projects (communities ask professionals to investigate a particular question) and *collegial* projects (non-professional individuals conduct largely independent research which may or may not be recognized by typical scientific authorities.

5. The integration of multiple disciplines into an approach based on the SES framework has been advocated by several authors, including recently Epstein et al. (2013) and Schlüter et al. (2014).
6. A modified version of the urban forests as social-ecological systems perspective is presented in Vogt et al. (in review) and on the BUFRG webpage: http://www.indiana.edu/~cipec/research/bufrg_about.php).
7. However, few of these components have been empirically evaluated to determine their impact on urban forest outcomes (but see Kenney et al. 2011).
8. Re-inventorying trees during the establishment and semi-mature phases between approximately 2 and 10 years after planting means that data collection could be combined with any remaining young tree maintenance (mulching, stake removal, training pruning, etc.).
9. This variable was collected at the suggestion of employees of Keep Indianapolis Beautiful, Inc., who teach volunteers to plant trees at the correct depth by maintaining the root flare.

REFERENCES

1. Anella, L., Hennessey, T. C., and Lorenzi, E. M. (2008). Growth of balled-and-burlapped versus bare-root Trees in Oklahoma , U.S. Arboriculture & Urban Forestry, 34(3), 200–203.
2. Arnold, C. L., & Gibbons, C. J. (1996). Impervious surface coverage: The emergence of a key environmental indicator. Journal of the American Planning Association, 62(2), 243–258. doi:10.1080/01944369608975688
3. Austin, M. E. (2002). Partnership opportunities in neighborhood tree planting initiatives: Building from local knowledge. Journal of Arboriculture, 28(4), 178–186.
4. Bancks, N. (2014). Count us in! Accuracy of volunteers in inventory (survey) data collection. 2014 Minnesota Shade Tree Short Course, 18-19 March 2014, St. Paul, MN.
5. Berrang, P., Karnosky, D. F., and Stanton, B. J. (1985). Environmental factors affecting tree health in New York City. Journal of Arboriculture, 11(6), 185–189.
6. Bloniarz, D. V, & Ryan, H. D. P. (1996). The use of volunteer initiatives in conducting urban forest resource inventories. Journal of Arboriculture, 22(2), 75–82.
7. Bonney, R., Shirk, J. L., Phillips, T. B., Wiggins, A., Ballard, H. L., Miller-Rushing, A. J., and Parrish, J. K. (2014). Next steps for citizen science. Science, 343, 1436–1437.
8. Boyce, S. (2010). It Takes a Stewardship Village: Effect of Volunteer Tree Stewardship on Urban Street Tree Mortality Rates, Cities and the Environment, 3(1), 1–8.
9. Buckstrup, M. J., and Bassuk, N. L. (2000). Transplanting success of balled-and-burlapped versus bare-root trees in the urban landscape. Journal of Arboriculture, 26(6), 298–308.
10. Clark, J. R., and Matheny, N. (1991). Management of mature trees. Journal of Arboriculture, 17(7), 173–184. doi:10.1016/0006-3207(92)90962-M

11. Clark, J. R., Matheny, N. P., Cross, G., and Wake, V. (1997). A model of urban forest sustainability. Journal of Arboriculture, 23(1), 17–30.
12. Clarke, K. C. (2009). A citizen science campaign encouraging urban forest professionals to engage the public in the collection of tree phenological data. American Geophysical Union Fall Meeting, 5-9 December 2014, San Francisco, CA.
13. Costello, L. (2013). Urban trees and water: An overview of studies on irrigation needs in the Western United States and a discussion regarding future research. Arboriculture & Urban Forestry, 39(3), 132–135.
14. Cowett, F. D., Associate, P., and Bassuk, N. L. (2012). SWAT (Student Weekend Arborist Team): A model for land grant institutions and cooperative extension systems to conduct street tree inventories. Journal of Extension, 50(3), art. 3FEA9.
15. Cox, M., Arnold, G., and Villamayor Tomas, S. (2010). A review of design principles for community-based natural resource. Ecology and Society, 15(4), art. 38. http://www.ecologyandsociety.org/vol15/iss4/art38/
16. Dahmus, M. E., and Nelson, K. C. (2013). Yard stories: Examining residents' conceptions of their yards as part of the urban ecosystem in Minnesota. Urban Ecosystems, 17(1), 173–194. doi:10.1007/s11252-013-0306-3
17. Day, S., Wiseman, P., Dickinson, S., and Harris, J. R. (2010). Tree root ecology in the urban environment and implications for a sustainable rhizosphere. Arboriculture & Urban Forestry, 36(4), 193–204.
18. Dickinson, J. L., Shirk, J., Bonter, D., Bonney, R., Crain, R. L., Martin, J., Phillips, T., and Purcell, K. (2012). The current state of citizen science as a tool for ecological research and public engagement. Frontiers in Ecology and the Environment, 10(6), 291–297. doi:10.1890/110236
19. Epstein, G., Vogt, J. M., Mincey, S. K., Cox, M., and Fischer, B. (2013). Missing ecology: integrating ecological perspectives with the social-ecological system framework. International Journal of the Commons, 7(2), 432–453. http://www.thecommonsjournal.org/index.php/ijc/article/view/371/331.
20. Evans, P. S., and Klett, J. E. (1985). The effects of dormant branch thinning on total leaf, shoot, and root production from bare-root Prunus cerasifera "Newportii." Journal of Arboriculture, 11(5), 149–151.
21. Gilman, E. F. (2001). Effect of nursery production method, irrigation, and inoculation with mycorrhizae-forming fungi on establishment of Quercus virginiana. Journal of Arboriculture, 27(1), 30–39.
22. Gilman, E. F. (2004). Effects of Amendments, soil additives, and irrigation on tree survival and growth. Journal of Arboriculture, 30(5), 301–310.
23. Gilman, E. F., and Beeson Jr., R. C. (1996). Production Method Affects Tree Establishment in the Landscape. Journal of Environmental Horticulture, 14(2), 81–87.
24. Grabosky, J., and Gilman, E. (2004). Measurement and prediction of tree growth reduction from tree planting space design in established parking lots. Journal of Arboriculture, 30(3), 154–164.
25. Graves, W. R., Dana, M. N., and Joly, R. J. (1989). Influence of root-zone temperature on growth of Ailanthus altissima (Mill.) Swingle. Journal of Environmental Horticulture, 7(2), 79–82.
26. Grove, J. M., Cadenasso, M., Burch, W., Pickett, S., Schwarz, K., O'Neil-Dunne, J., Wilson, M., Troy, A., and Boone, C. (2006). Data and methods comparing social

structure and vegetation structure of urban neighborhoods in Baltimore, Maryland. Society & Natural Resources, 19(2), 117–136. doi:10.1080/08941920500394501
27. Holling, C.S. (1996). Surprise for science, resilience for ecosystems, and incentives for people. Ecological Applications, 6(3), 733-735.
28. Iakovoglou, V., Thompson, J., and Burras, L. (2002). Characteristics of trees according to community population level and by land use in the U.S. Midwest. Journal of Arboriculture, 28(2), 59–69.
29. Iakovoglou, V., Thompson, J., Burras, L., and Kipper, R. (2001). Factors related to tree growth across urban-rural gradients in the Midwest, USA. Urban Ecosystems, 5, 71–85.
30. International Tree Failure Database (ITFD). 2014. http://svinetfc8.fs.fed.us/natfdb/. Accessed 18 April 2014.
31. International Union of Forest Research Organizations, International Society of Arboriculture, United States Forest Service, & Urban Natural Resources Institute. (2010). Standards for Urban Forestry Data Collection: A Field Guide, Draft 2.0. Forestry (2.0 ed., p. 29). http://www.unri.org/standards/wp-content/uploads/2010/08/Version-2.0-082010.pdf. Accessed 18 April 2014.
32. i-Tree. 2014. http://www.itreetools.org. Accessed 18 April 2014.
33. Jack-Scott, E., Piana, M., Troxel, B., Murphy-Dunning, C., and Ashton, M. S. (2013). Stewardship success: How community group dynamics affect urban street tree survival and growth. Arboriculture & Urban Forestry, 39(4), 189–196.
34. Jutras, P., Prasher, S. O., and Mehuys, G. R. (2009). Prediction of street tree morphological parameters using artificial neural networks. Computers and Electronics in Agriculture, 67(1-2), 9–17. doi:10.1016/j.compag.2009.02.008
35. Kenney, W. A., van Wassenaer, P. J. E., and Satel, A. L. (2011). Criteria and indicators for strategic urban forest planning and management. Arboriculture & Urban Forestry, 37(3), 108–117.
36. Koeser, A., Hauer, R., Norris, K., and Krouse, R. (2013). Factors influencing long-term street tree survival in Milwaukee, WI, USA. Urban Forestry & Urban Greening, 12(4), 562-568. doi:10.1016/j.ufug.2013.05.006
37. Kozlowski, T. T., and Pallardy, S. G. (1997). Growth Control in Woody Plants. San Diego, CA: Academic Press, Inc.
38. Kramer, P. (1987). The role of water stress in tree growth. Journal of Arboriculture, 13(2), 33–38.
39. Krizek, D. T., and Dubik, S. P. (1987). Influence of water stress and restricted root volume on growth and development of urban trees. Journal of Arboriculture, 13(2), 47–55.
40. Lambert, B. B., Harper, S. J., and Robinson, S. D. (2010). Effect of container size at time of planting on tree growth rates for baldcypress (Taxodium distichum (L.) Rich), red maple (Acer rubrum L.), and longleaf pine (Pinus palustris Mill.). Arboriculture & Urban Forestry, 36(2), 93–99.
41. Larson, K. L., Hall, S. J., Cook, E. M., Funke, B., Strawhacker, C. A., and Turner, V. K. (2008). Social-ecological dynamics of residential landscapes: Human drivers of management practices and ecological structure in an urban ecosystem context. Tempe, AZ: Central Arizona Phoenix Long-Term Ecological Research. http://

caplter.asu.edu/docs/papers/2008/CAPLTER/Larson_etal_2008.pdf. Accessed 18 April 2014.
42. Leibowitz, R. (2012). Urban tree growth and longevity: An international meeting and research symposium white paper, Arboriculture & Urban Forestry, 38(5), 237–241.
43. Liu, J., Dietz, T., Carpenter, S. R., Alberti, M., Folke, C., Moran, E., Pell, A. N., Deadman, P., Kratz, T., Lubchenco, J., Ostrom, E., Ouyang, Z., Provencher, W., Redman, C. L., Schneider, S. H., and Taylor, W. W. (2007). Complexity of coupled human and natural systems. Science, 317(5844), 1513–6. doi:10.1126/science.1144004
44. Lu, J. W. T., Svendsen, E. S., Campbell, L. K., Greenfeld, J., Braden, J., King, K. L., and Falxa-Raymond, N. (2011). Biological, social, and urban design factors affecting young street tree mortality in New York City. Cities and the Environment, 3(1), 1–15. http://digitalcommons.lmu.edu/cate/vol3/iss1/5/
45. Morton Arboretum Soil Science (MASS) Laboratory. 2014. Morton Arboretum Soil Science. http://www.masslaboratory.org/urban-site-index.html. Accessed 18 April 2014.
46. Miller, R. W., and Sylvester, W. A. (1981). An economic evaluation of the pruning cycle. Journal of Arboriculture, 7(4), 109–112.
47. Miller-Rushing, A., Primack, R., and Bonney, R. (2012). The history of public participation in ecological research. Frontiers in Ecology and the Environment, 10(6), 285–290. doi:10.1890/110278
48. Mincey, S. K., Hutten, M., Fischer, B. C., Evans, T. P., Stewart, S. I., and Vogt, J. M. (2013). Structuring institutional analysis for urban ecosystems: A key to sustainable urban forest management. Urban Ecosystems, 16(3), 553–571. doi:10.1007/s11252-013-0286-3
49. Mincey, S. K., and Vogt, J. M. (2014). Watering strategy, collective action, and neighborhoodplanted trees: A case study of Indianapolis, Indiana, U.S. Arboriculture & Urban Forestry, 40(2), 84–95.
50. Nassauer, J. I., Wang, Z., and Dayrell, E. (2009). What will the neighbors think? Cultural norms and ecological design. Landscape and Urban Planning, 92(3-4), 282–292. doi:10.1016/j.landurbplan.2009.05.010
51. National Audubon Society. 2014. Christmas Bird Count. http://birds.audubon.org/christmasbird-count. Accessed 18 April 2014.
52. Nielsen, A. B., Östberg, J., and Delshammar, T. (2014). Review of urban tree inventory methods used to collect data at single-tree level. Arboriculture & Urban Forestry, 40(2), 96–111.
53. Nowak, D. J., McBride, J. R., and Beatty, R. A. (1990). Newly planted street tree growth and mortality. Journal of Arboriculture, 16(5), 124–129.
54. NYC Parks. 2014. Young Street Tree Mortality. http://www.nycgovparks.org/trees/ystm. Accessed 18 April 2014.
55. NYC Parks, USDA Forest Service Northern Research Station, Rutgers University, & Parsons, T. N. S. F. D. (2010). New York City's Young Street Tree Mortality Study Site Assessment Tools Description. October. New York City, NY. Retrieved from http://www.nycgovparks.org/sub_your_park/trees_greenstreets/images/how_to_assess_your_citys_st_survival_10-6-2010.pdf
56. Ostrom, E. (1990). Governing the Commons. Cambridge, UK: Cambridge University Press.

57. Ostrom, E. (2005). Understanding Institutional Diversity. Princeton, NJ: Princeton University Press.
58. Ostrom, E. (2009). A general framework for analyzing sustainability of social-ecological systems. Science, 325(5939), 419–422. doi:10.1126/science.1172133
59. Ostrom, E., and Cox, M. (2010). Moving beyond panaceas: A multi-tiered diagnostic approach for social-ecological analysis. Environmental Conservation, 37(04), 451–463. doi:10.1017/S0376892910000834
60. Peper, P. P., and McPherson, E. G. (2003). Evaluation of four methods for estimating leaf area of isolated trees. Urban Forestry & Urban Greening, 2, 19–29.
61. Rhoades, R. W., and Stipes, R. J. (1999). Growth of trees on the Virginia Tech campus in response to various factors. Journal of Arboriculture, 25(4), 211–217.
62. Roman, L. A. (2013). Urban tree mortality. Dissertation. University of California, Berkeley.
63. Roman, L. A., McPherson, E. G., Scharenbroch, B. C., and Bartens, J. (2013). Identifying common practices and challenges for local urban tree monitoring programs across the United States, Arboriculture & Urban Forestry, 39(6), 292–299.
64. Roman, L. A., and Scatena, F. N. (2011). Street tree survival rates: Meta-analysis of previous studies and application to a field survey in Philadelphia, PA, USA. Urban Forestry & Urban Greening, 10(4), 269–274. doi:10.1016/j.ufug.2011.05.008
65. Scharenbroch, B. C., Lloyd, J. E., and Johnson-Maynard, J. L. (2005). Distinguishing urban soils with physical, chemical, and biological properties. Pedobiologia, 49, 283–296. doi:10.1016/j.pedobi.2004.12.002
66. Schlüter, M., Hinkel, J., Bots, P., and Arlinghaus, R. (2014). Application of the SES framework for model-based analysis of the dynamics of social-ecological systems. Ecology & Society, 19(1), art. 36. http://www.ecologyandsociety.org/vol19/iss1/art36/.
67. Shirk, J. L., H. L. Ballard, C. C. Wilderman, T. Phillips, A. Wiggins, R. Jordan, E. McCallie, M. Minarchek, B. V. Lewenstein, M. E. Krasny, and R. Bonney. (2012). Public participation in scientific research: a framework for deliberate design. Ecology and Society, 17(2), 29. http://dx.doi.org/10.5751/ES-04705-170229
68. Smith, K. D., May, P. B., and Moore, G. M. (2001). The influence of compaction and soil strength on the establishment of four Australian landscape trees. Journal of Arboriculture, 27(1), 1–7.
69. Solfjeld, I., and Hansen, O. B. (2004). Post-transplant growth of five deciduous Nordic tree species as affected by transplanting date and root pruning. Urban Forestry & Urban Greening, 2, 129–137.
70. Stoffberg, G. H., van Rooyen, M. W., van der Linde, M. J., and Groeneveld, H. T. (2008). Predicting the growth in tree height and crown size of three street tree species in the City of Tshwane, South Africa. Urban Forestry & Urban Greening, 7(4), 259–264. doi:10.1016/j.ufug.2008.05.002
71. Struve, D. K., Burchfield, L., and Maupin, C. (2000). Survival and growth of transplanted largeand small-caliper red oaks. Journal of Arboriculture, 26(3), 162–169.
72. Symes, P., and Connellan, G. (2013). Water management strategies for urban trees in dry environments: Lessons for the future. Arboriculture & Urban Forestry, 39(3), 116–124.

73. Tattar, T. A. (1980). Non-infectious diseases of trees. Journal of Arboriculture, 6(1), 1-4.
74. Tretheway, R., Simon, M., McPherson, E. G., and Mathis, S. (1999). Volunteer-Based Urban Forest Inventory and Monitoring Programs: Results from a Two-Day Workshop, 17-18 February 1999, Sacramento, CA. http://www.fs.fed.us/psw/programs/uesd/uep/products/5/cufr_90.pdf. Accessed 18 April 2014.
75. United States Environmental Protection Agency (US EPA). (2008). Reducing Urban Heat Islands: Compendium of Strategies: Urban Heat Island Basics. http://www.epa.gov/heatisland/resources/compendium.htm. Accessed 18 April 2014.
76. Vogt J. M., Hauer, R. J., and Fischer, B. C. (In review). The costs of maintaining and not maintaining trees: A review of the urban forestry and arboriculture literature.
77. Vogt, J. M., Mincey, S. K., Fischer, B. C., and Patterson, M. (2013). Planted Tree Re-inventory Protocol, Version 1.0. Bloomington, IN: Bloomington Urban Forestry Research Group at the Center for the study of Institutions, Population and Environmental Change, Indiana University. http://www.indiana.edu/~cipec/research/bufrg_protocol Accessed 18 April 2014.
78. Vogt, J. M., Watkins, S. L., Mincey, S. K., Patterson, M., and Fischer, B. C. (In review). Explaining planted-tree survival and growth in urban neighborhoods: A social-ecological approach to studying recently-planted trees in Indianapolis.
79. Vogt, J. M., Wilson, M., and Swilik, J. (2012). Engaging high school students in urban tree growth research. Conference on Public Participation in Scientific Research at the Ecological Society of American Annual Meeting, 3-4 August 2012, Portland, OR.
80. University of Minnesota (UM) Department of Forest Resources, UM Extension Service, and Minnesota Department of Agriculture. (2014). Urban & Community Forestry: Community Engagement and Preparedness Program. http://www.mntreesource.com. Accessed 18 April 2014.
81. Urban Forest Map, Pennsylvania Horticultural Society, Philadelphia Parks and Recreation, Township of Lower Merion, Azavea, and Delaware Valley Regional Planning Commission. (2014). PhillyTreeMap Beta. http://www.phillytreemap.org. Accessed 18 April 2014.
82. Urban Tree Growth and Longevity (UTGL) Working Group. (2014a). Urban Tree Growth and Longevity. http://www.urbantreegrowth.org. Accessed 18 April 2014.
83. Urban Tree Growth and Longevity (UTGL) Working Group. (2014b). Urban Tree Monitoring Protocol. http://www.urbantreegrowth.org/urban-tree-monitoring-protocol.html. Accessed 18 April 2014.
84. Whitcomb, C. E. (1979). Factors affecting the establishment of urban trees. Journal of Arboriculture, 5(10), 217–219.
85. Zooniverse. 2014. GalaxyZoo. http://www.galaxyzoo.org. Accessed 18 April 2014.

The entire Planted Tree Re-Inventory Protocol is available online at http://www.indiana.edu/~cipec/research/bufrg_protocol.php.

PART IV

IMPROVING OUR UNDERSTANDING OF URBAN FORESTS

CHAPTER 7

110 Years of Change In Urban Tree Stocks and Associated Carbon Storage

DANIEL F. DÍAZ-PORRAS, KEVIN J. GASTON, AND KARL L. EVANS

7.1 INTRODUCTION

It is important to document temporal changes in urban green space and its associated vegetation, because of the rapidly expanding and dynamic nature of urban areas, and the key role of this vegetation in supporting urban biodiversity and providing ecosystem services (Seto et al. 2012; Gaston et al. 2013). Trees, particularly large ones, are keystone structures in many ecosystems, including urban areas (Lindenmayer et al. 2012; Stagoll et al. 2012). In towns and cities, the abundance and nature of trees plays a major role in determining the structure and composition of faunal assemblages (Evans et al. 2009; Stagoll et al. 2012). Trees and shrubs also play a key role in providing ecosystem services in urban areas, primarily because they comprise a considerable proportion of the vegetation's biomass (Davies et al. 2011; Roy et al. 2012). These benefits include a range of cultural services and improvements to human health and well-being (Ulrich 1986; Kuo and Sullivan 2001; Maas et al. 2006; Fuller et al. 2007). Urban vegetation also provides several regulating services including reducing air pollution (Donovan et al. 2005), the urban heat island effect (Lindberg and Grimmond 2011; Hall et al. 2012), noise pollution (Islam et al. 2012), and flood risk (Stovin et al. 2008). Finally, urban trees make a significant contribution to carbon sequestration (Nowak and Crane 2002).

Urban trees have historically faced a number of threats, and will continue to do so. Heat and drought stress seem likely to be amplified in urban areas due to the urban heat island effect, reduced water infiltration into soils due to the dominance of impervious surfaces, and soil compaction (Sieghardt et al. 2005). The urban heat island effect can also contribute to increased susceptibility of urban tree to pests (Meineke et al. 2013). Urban trees may also suffer more from pests and exotic diseases than their rural counterparts due to increased exposure to horticultural trade, for example, Asian long-horned beetle *Anoplophora glabripennis* became established in North America in urban areas and has only recently invaded rural ones (Dodds and Orwig 2011). Whilst air pollution can reduce growth rates of urban trees, there are some examples of increased growth rates in response to higher CO_2 concentrations in urban areas (Evans 2010). Finally, urban trees are also more likely to be prevented from reaching their full growth potential due to the association between height and the probability of damaging urban infrastructure or blocking light.

Empirical data assessing changes in the nature and composition of urban green space are typically limited to use of remote-sensing data (e.g., Pauleit et al. 2005; Dallimer et al. 2011; Gillespie et al. 2012). Due to the timing of the development of appropriate technologies, such studies are inevitably restricted to a few recent decades; this is a small time period relative to the age of many urban areas, and assessments over longer-time periods are essential to provide a complete understanding of the impacts of urbanization. In addition, remote-sensing technologies have not always had sufficient capacity to distinguish individual components of green space, such as trees and shrubs, or to record their size. Given the strong relationship between ecosystem service provision and vegetation biomass and thus tree size (see above), this further limits assessment of the dynamics of urban vegetation. Collections of historical photographs provide a valuable source of detailed data on past environmental conditions that can be used to track long-term environmental change, which overcomes these limitations (Pennisi 2013). This approach is time-consuming as it requires finding a large number of dated historical images that include the key items of interest, and then refinding the original location from which these images were taken. Repeat photography has great value, however, and has been used to assess rates of glacial retreat, and changes in plant growth rates, vegetation composition, and forest cover (Chen et al. 2011; Myers-Smith et al. 2011; Van Bogaert et al. 2011). Such studies have rarely focused on urban areas, although Nowak (1993) used historical photographs in combination with other historical documents to assess vegetation change in Oakland, California. Monge-Nájera and Pérez-Gómez (2010) also used repeat photography to assess change in tree cover in San Jose, Costa Rica, but could only find nine suitable historical images.

Here, we employ repeat photography to assess long-term changes in the number and size of trees over a 110-year period using Sheffield, the fifth largest urban area (c. 555,500 people; Office for National Statistics 2010) in the UK, as a case study. We then use these data to assess temporal change in the contribution of the urban tree stock to aboveground carbon storage. We also test whether the temporal dynamics in the stock of urban trees is uniform across the urbanized region, or varies with the intensity of urban development. This is important because urban areas are not homogenous (Davies et al. 2008), and the magnitude and intensity of change can vary with urban form.

7.2 METHODS

7.2.1 OBTAINING AND REPEATING HISTORICAL PHOTOGRAPHS

We used a paired design and compared photographs taken in the 1900s and 1950s with those taken in 2010, although the two sets of historical images were not taken in the same location. Our objective was to calculate broad trends in the numbers of trees of different size categories to generate an index of change in urban tree stocks. Urban Sheffield was defined as those

1 × 1 km squares with at least 25% hard surface. Historical photographs were obtained from Sheffield's Local Studies Library online database (http://www.picturesheffield.com), which contains approximately 35,000 images, primarily from the 1900s. All images taken between 1900 and 1909 (referred to as the 1900s) or between 1950 and 1959 (referred to as the 1950s) were selected. The 1900s is the earliest decade for which sufficient images were available, and the 1950s represents a period of intense urban development following the Second World War.

We consider the set of historical photographs to represent an unbiased haphazard sampling design that is sufficient for estimating general trends in the urban tree stock for three reasons. First, the original photographic locations seem highly unlikely to have been selected on the basis of their tree cover. This is because the primary reason for taking the photographs was to record people or buildings—often both (e.g., photos of people taken outside their homes or work places). The massive variation in tree cover recorded in the historical images is one indication that positive or negative biases toward including trees in the historical images are unlikely to be large. Second, the locations of the historical images cover much of the focal urban region of Sheffield, albeit with an inevitable concentration in older urban areas that were urbanized in the 1900s and 1950s, and represent the full range of variation in urban form as assessed by the amount of green space currently present in the area (Fig. 1; and see Results). Finally, it seems unlikely that the location of the historical images would be biased according to future trends in tree cover as these were unknown at the time the images were taken.

Aerial images and those that mainly comprised the inside of buildings or obscured views were excluded. The potential to obtain a current image at precisely the same location as the historical image was assessed using the street view tool of Google Earth using three criteria: (1) the ability to use features in the historical image to pinpoint its exact location, (2) that the historic landscape captured in the original image was not currently obscured, and (3) that the site was accessible. When the potential could not be assessed using the street view tool (e.g., inside large parks), a site visit was conducted. Following these processes, 121 and 109 images were selected for the 1900s and 1950s, respectively.

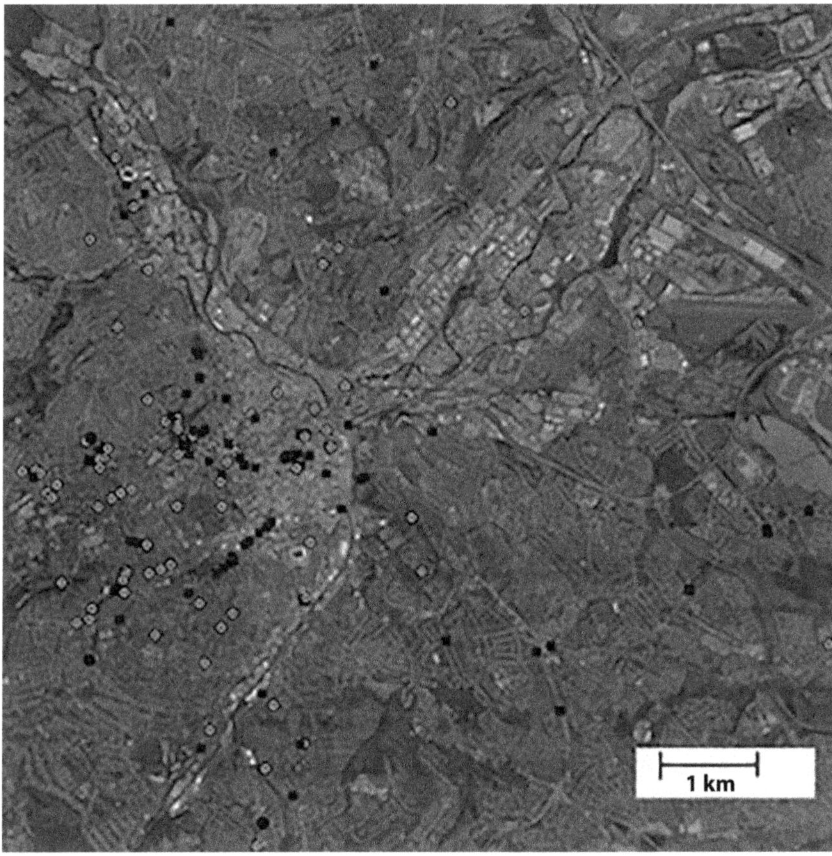

FIGURE 1: The location of the historical photographs from the 1900s (white circles) and the 1950s (blue squares) in urban Sheffield. Base imagery is from Google Earth and comprises a composite of images taken in 2008 and 2011.

Additional searches were made for images from unrepresented boroughs taken during the contiguous decades, that is, within the 1890s and 1910s for the 1900s, and within the 1940s and 1960s for the 1950s. This resulted in a selection of 17 and 24 additional photographs, respectively, for 1890–1919 and 1940–1969. The former is hereafter referred to as the 1900s (88% of images are from 1900–1909) and the latter as the 1950s (82% of images are from 1950–1959).

Fieldwork was carried out from June to early September 2010. Repeat photographs were taken using a 4.6× optical zoom digital camera (12.2 megapixels) and matched the position and direction of historical photographs as closely as possible. Each photographic location was georeferenced using a GPS. About 61 of the 271 historical photographs could not be repeated due to a failure to find the precise location of the original image or because the precise historical view could not be reconstructed. This left 106 pairs comparing the 1900s with 2010, and 104 pairs comparing the 1950s with 2010.

7.2.2 QUANTIFYING CHANGES IN THE TREE STOCK

All shrubs and trees present in the entire photograph were identified using the following height categories: (1) <2 m, (2) 2–5 m, (3) 5–10 m, and (4) >10 m. This was achieved by comparing the heights of individual trees and shrubs, by eye, with standardized reference heights of other features typically present in the urban landscape that were measured in the field; in addition, people were assumed to be <2 m tall. Whilst use of these reference heights does not provide a precise measure of the height of focal trees or shrubs, it provides an unbiased mechanism that can be applied to both historical and current time periods with which each shrub/tree can be accurately placed within a height category.

Aboveground dry-weight tree biomass was calculated using the allometric equation from Davies et al. (2011): biomass (kg) = 0.566*(height in meters)2.315, and summing across the total number of trees in each height category. When our height categories were bounded, we used their midpoint as an estimate of tree height, for the unbounded category of trees >10 m, and we repeated calculations using a range of tree height estimates (12, 15, and 18 m) that cover the full range of plausible midpoints based on observed size distributions of urban trees in the U.K. (Davies et al. 2011). The allometric equation that we used was developed for broad-leaved trees in urban Leicester, located 90 km south of Sheffield and of similar urban form. This equation takes into account the relative abundance of different tree species, and uses species, genus, or family-specific allometric relationships. This approach was adopted as historical photographs were rarely of sufficient quality to allow trees to be identified to species or genus. This will reduce the precision of our estimates of tree biomass as there may be some shifts in composition of the tree assemblage across time periods, but it does not prevent us from generating sufficiently accurate estimates to calculate overall trends in tree biomass and resultant carbon storage. This is because the form of allometric equations is fairly similar across different broad-leaved tree species, and broad-leaved trees comprised the vast majority of shrubs and trees in the historical and repeated images. This concurs with the regional and national pattern (Britt and Johnston 2008), and additional data collected as part of biodiversity surveys in Sheffield, that found that broad-leaved trees comprised 92.8% of trees. The five commonest tree species were sycamore *Acer pseudoplatanus*, ash *Fraxinus excelsior*, pedunculate oak *Quercus robur*, silver birch *Betula pendula*, and cherry *Prunus* spp. These data were obtained in 2010 from 140 sampling points selected using a random stratified design with regard to the amount of green space as described by Bonnington et al. (in press). We thus consider that our calculations provide a reasonably robust estimate of relative temporal change in tree biomass. Aboveground tree biomass (kg) was transformed to a carbon storage figure using the broadleaf conversion factor of 0.48 (Milne and Brown 1997).

7.2.3 CALCULATING THE PERCENTAGE OF GREEN SPACE

We wished to assess how trends in urban tree cover varied across different urban forms, which is most frequently measured by the amount of green space, or its inverse the amount of hard surface present in a given area. To achieve this, the amount of green space (i.e., vegetated surface, the majority of which is grass) currently present in the 250 × 250-m grid cell surrounding each photographic location was calculated using an OS Master 1:10000 scale Georeferenced TIFF raster map for the 2005–2009 period obtained from the Digimap Ordnance Survey Collection (via http://edina.ac.uk).

7.2.4 STATISTICAL ANALYSES

All analyses were conducted using SPSS 16 (SPSS Inc., Chicago, IL) or SAS vs 9.3 (SAS Institute Inc., Cary, NC). We have two sets of paired photographs (1900s and 2010; 1950s and 2010), and the primary focus was to exploit this paired experimental design. We thus used a matched paired t-test to compare the urban tree stock (total number of trees, and numbers in each height category) that was present in the 1900s with that present in 2010, and to compare the tree stock in the 1950s with that present in 2010 (data on differences in the number of trees did not differ from a normal distribution; Kolmogorov–Smirnov test, $P > 0.05$ in all cases). Photographic locations were different in the 1900s and 1950s and thus do not involve a paired design, and differences in the number of trees in these time periods did not follow a normal distribution. Changes in the urban tree stock between the 1900s and 1950s were thus analyzed using Mann–Whitney U-tests.

The percentage change in the number of shrubs/trees was calculated (for the total number of trees and for each height category except for trees > 10 m, see below) by adding one to the number of trees present to enable percentages to be calculated at sites with no trees. Percentages were then square-root transformed to meet statistical assumptions of normality; transformations were conducted on absolute values, and following transformation values that were originally negative were multiplied by minus one to preserve their original sign. We then used general linear models to model the transformed percentage change in the number of shrubs/trees as a function of the percentage of green space currently present in the surrounding 250 × 250-m grid cell. We did so using general linear models that include both linear and square terms as predictors, but removed the square term from the final model unless it was statistically significant ($P < 0.05$). When the square term was included in the final model, we conducted a break point regression to assess the nature of the relationship between the percentage increase in shrubs/trees and green space below and above the turning point of the quadratic model. Moran's I values were consistently very low (<0.01 for all response variables) indicating that the data contained negligible spatial autocorrelation.

7.3 RESULTS

7.3.1 1900S–2010

The total number of shrubs/trees increased by 50.5% ($t = 6.20$, $df = 105$, $P < 0.001$; $df = 105$ in all cases; Fig. 2). Most size categories also exhibited significant increases: <2 m (67.6%, $t = 4.06$, $P = 0.0001$), 5–10 m (33.4%, $t = 2.01$, $P = 0.05$), >10 m (214.7%, $t = 3.36$, $P = 0.0001$), but the 13.7% increase in the number of shrubs/trees between 2–5 m was not significant ($P = 0.39$; Fig. 2). Aboveground carbon storage in trees approximately doubled from the 1900s–2010, with the rate of increase being little influenced by the choice of midpoint for the unbounded height category (i.e., trees > 10 m; Table 1A).

7.3.2 1950S–2010

The total number of shrubs/trees increased by 95.8% ($t = 6.91$, $df = 103$, $P < 0.001$; $df = 103$ in all cases; Fig. 2). Most size categories also exhibited significant increases: <2 m (65.8%, $t = 3.05$, $P = 0.003$), trees between 2–5 m (88.8%, $t = 4.12$, $P = 0.001$), trees between 5–10 m (151.2%, $t = 7.24$, $P = 0.001$); the 52.3% increase in the number of trees > 10 m was not significant ($P = 0.30$; Fig. 2). From the 1950s–2010, aboveground carbon storage in trees approximately doubled, with the choice of midpoint for the unbounded height category again having little influence on the estimated rate of change (Table 1B).

7.3.3 1900S–1950S

The total number of shrubs/trees declined by 37.5% ($U = 4416.0$, $P = 0.01$). The numbers of shrubs/trees in each of the height categories also tended to decline during this period, but these differences were only significant for trees between 2–5 m in height (53.2%, $U = 4066$, $P < 0.001$), with other differences not being significant: <2 m (23.7%, $U = 5079$, $P = 0.295$), 5–10 m (35.1%, $U = 4942$, $P = 0.175$), and >10 m (53.7%, $U = 5083$, $P = 0.119$).

Table 1. Change in aboveground carbon storage of the urban tree stock in Sheffield (U.K.) from (A) 1900 to 2010, and (b) 1950 to 2010. Biomass is calculated using the allometric equation for broad-leaved trees in urban Leicester (U.K.) from Davies et al. (2011) and converted to carbon storage following Milne and Brown (1997). Data are calculated using the summed number of trees present in historical and repeated photographs in four height categories (<2 m, 2–5 m, 5–10 m, > 10 m), and using the midpoint of each height category. Ratios of change are broadly consistent regardless of the midpoint used for the largest unbounded height category.

Height midpoint used for trees > 10 m	Aboveground tree carbon (kg) 1900	Aboveground tree carbon (kg) 2010	Carbon ratio (2010:1900)
(A)			
12 m	18142.8	35426.6	1.95
15 m	22599.2	48853.5	2.16
18 m	28399.6	66330.1	2.34
(B)			
12 m	13663.1	29804.4	2.18
15 m	16209.6	33682.1	2.08
18 m	19524.1	38729.2	1.98

7.3.4 RELATIVE ABUNDANCE BY HEIGHT CLASS IN 2010

Pooling data from both sets of locations of historical images revealed that, across the 3598 trees captured, 36% were <2 m tall, 22% were 2–5 m tall, 34% were 5–10 m tall, and 8% were greater than 10 m in height.

7.3.5 RELATIONSHIPS BETWEEN CHANGES IN TREE STOCKS AND AMOUNT OF GREEN SPACE

Between the 1900s and 2010, the percentage increase in the total number of shrubs and trees and of trees between 5 m and 10 m tall was negatively associated with the amount of green space in the surrounding 250 × 250-m grid cells (Fig. 3A,B; Table 2A). The percentage increase in shrubs/trees that were <2 m and 2–5 m tall exhibited the same trend, but this was not

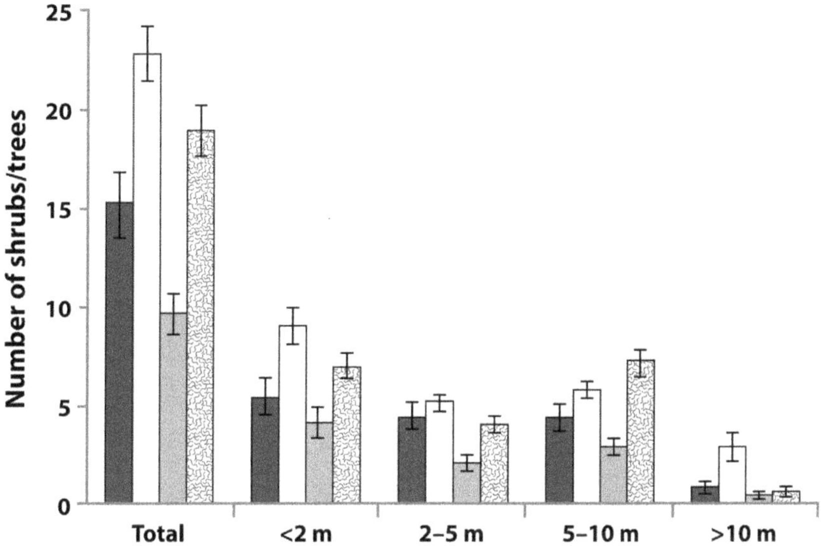

FIGURE 2: The number of shrubs and trees in urban Sheffield present in the 1900s (dark grey bars), 1950s (pale grey bars), and 2010 (white bars). Data are from 106 paired repeat photographs taken in the 1900s and 2010 (left-hand white bar in each category), and 104 paired repeat photographs taken in the 1950s and 2010 (right-hand white bar). Error bars represent standard errors.

statistically significant (Table 2). In contrast, the percentage increase in the number of trees that were taller than 10 m was greatest in areas that currently contained the most green space (Table 2A; Fig. 3C). Between the 1950s and 2010, the percentage increase in shrubs/trees that were <2 m tall exhibited a unimodal relationship with green space (no other relationships were statistically significant; Table 2B). Using a break point regression around the turning point of this unimodal relationship revealed that there was a significant positive association between the percentage increase in shrubs/trees that were <2 m tall until green space exceeded c. 40% of the surrounding 250 × 250-m grid cell (r^2 = 15.5%; $F_{1,40}$ = 7.34, P = 0.01; parameter estimate 0.539 ± 0.200), after which the percentage increase in shrubs/trees was not associated with the amount of green space (r^2 = 0.015%; $F_{1,60}$ = 0.94, P = 0.34; parameter estimate −0.110 ± 0.114).

Table 2. Relationships between percentage change in tree stocks in urban Sheffield from (A) the 1900s–2010, and (B) the 1950s–2010 in repeated historical photos and the amount of current green space in the surrounding 250 Ч 250-m grid cell. The percentage change in tree stocks was square-root transformed prior to analysis. All data refer to linear terms unless otherwise indicated.

Height class	Model r^2, %	Parameter estimate (±SE)	F ratio; P value	Equation
(A)				
All trees	10.99	−0.189 ± 0.053	$F_{1,104}$ = 12.84; P = 0.0005	Y = 20.302 −0.189x
<2 m	0.25	−0.029 ± 0.056	$F_{1,104}$ = 0.26, P = 0.609	n/a
2–5 m	2.54	−0.081 ± 0.049	$F_{1,104}$ = 2.71, P = 0.103	n/a
5–10 m	6.32	−0.128 ± 0.048	$F_{1,104}$ = 7.02, P = 0.009	Y = 14.580−0.128x
>10 m	6.76	0.134 ± 0.049	$F_{1,104}$ = 7.54, P = 0.007	Y = −1.345 + 0.134x
(B)				
All trees	2.71	−0.103 ± 0.061	$F_{1,102}$ = 2.84; P = 0.095	n/a
<2 m	4.92	Linear term: 0.307 ± 0.175	Linear term $F_{1,101}$ = 0.79, P = 0.082;	Y = 8.831 + 0.307x - 0.004x^2
		Square term: −0.004 ± 0.002	Square term $F_{1,101}$ = 4.43, P = 0.034;	
2–5 m	0.67	−0.035 ± 0.043	$F_{1,102}$ = 0.69, P = 0.409	n/a
5–10 m	3.13	−0.083 ± 0.046	$F_{1,102}$ = 3.29, P = 0.073	n/a
>10 m	0.12	−0.009 ± 0.023	$F_{1,102}$ = 0.13, P = 0.718	n/a

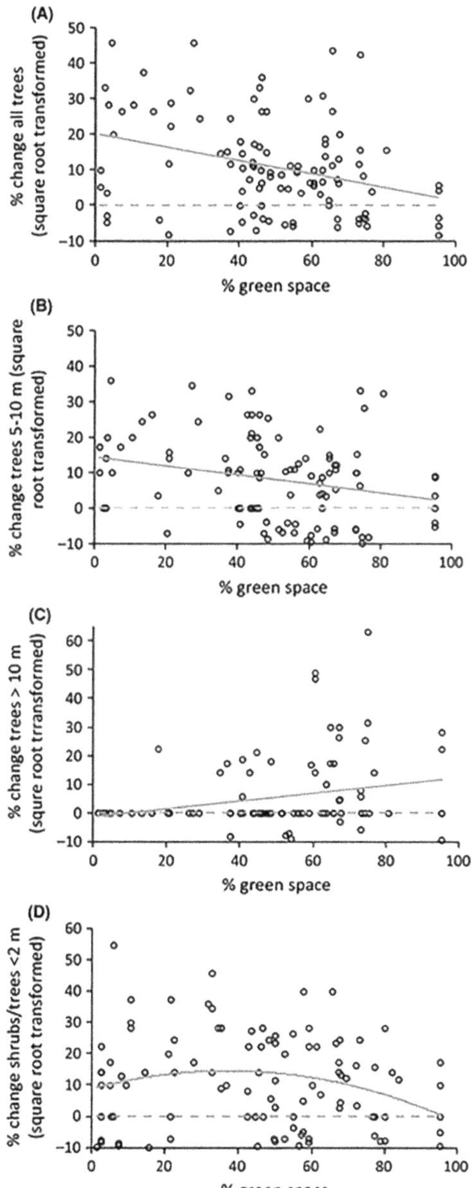

FIGURE 3: Relationships between the percentage increase in shrubs/trees and the amount of green space in the surrounding 250 Ч 250-m grid cell for (A) all shrubs/trees between the 1900s and 2010, (B) trees that are 5–10 m tall between the 1900s and 2010, (C) trees >10 m between the 1900s and 2010, and (D) trees <2 m between the 1950s and 2010.

7.4 DISCUSSION

We demonstrate that repeat photography can yield valuable data for long-term monitoring of urban tree stocks, and associated ecosystem services. Between the 1900s and 2010, shrubs/trees within urban Sheffield increased by over 50%. Equivalent studies conducted over comparable time periods are rare, and none have been conducted in regions with the long history of urbanization that characterizes our study, which further hinders direct comparisons. It is notable though that studies conducted in regions where forest cover is naturally limited, such as South-West North America, tend to find increased urban tree cover. In Oakland, California, for example, tree cover increased from approximately 5% during the city's initial development (1850s–1890s) to approximately 20% in 1991 (Nowak 1993). Similarly, tree densities more than doubled from the 1920s to the turn of the century at two urban sites near Los Angeles, California, although a small number of urban areas had decreased tree cover (Gillespie et al. 2012). In contrast, a 5% decrease in urban tree cover occurred from the 1890s–2010 in San José, Costa Rica (Monge-Nájera and Pérez-Gómez 2010): a region that naturally has a high level of forest cover.

The significant increase in the number of urban shrubs/trees in Sheffield since the 1900s is thus not unprecedented, but does represent one of the most marked rates of increase documented to date. One factor that may contribute to this is that in the early 1900s, past human activities had reduced tree cover across England to just 6%, and to less than 4% across Yorkshire, the county in which Sheffield is located (Forestry Commission 2001). The increase in total shrubs/trees was even more marked (c. 100%) from the 1950s–2010, due to a decrease in urban tree cover in the first half of the twentieth century which contrasts with a static trend in tree cover at the national level across this time period (Forestry Commission 2001). This decrease from 1900 to 1950 in urban tree abundance is likely to be a consequence of the marked urban intensification during this period, and bombing (and associated redevelopment) during the Second World War. The pattern that we find in Sheffield is similar to the initial trends in urban tree cover that arose in Baltimore, Maryland, with an initial decrease from 1914 to 1938, which was then followed by an increase till the 1970s (Zhou et al. 2011). There has subsequently been a decline in urban tree cover in Baltimore, resulting in no net change from 1914 to 2004. It should thus not be assumed that the increase in urban tree cover that we document in Sheffield will be maintained in the future, especially given the numerous and increasing threats to urban trees that seem likely to increase mortality rates (see Introduction).

We find clear evidence that small trees, that is, those less than two meters tall, are now commoner in urban Sheffield than they were in both the 1900s (68% increase) and 1950s (66% increase). Natural seedling abundance and establishment is lower in urban woodlands than rural ones, suggesting that natural regeneration is suppressed in urban areas (Oldfield et al. 2013). It thus seems likely that the increase in small trees since the 1900s and 1950s is at least partly driven by urban tree planting initiatives. Whilst explanatory power is somewhat limited, there is a tendency for smaller trees to exhibit larger increases in abundance in the areas with least green space, that is, the most intensively urbanized areas. This strengthens the conclusion that urban tree planting programmes have contributed to the increase in the number of small trees, as natural regeneration is likely to be particularly low in such sites.

The increase in the number of trees from the 1950s–2010 becomes larger as tree size increases from <2 m (66%), to 2–5 m (89%), and to 5–10 m (150%). There is insufficient data on the annual height increments of broad-leaved trees in urban environments to estimate robustly the age of these trees. Growth rates of *Prunus*, *Acer*, and *Quercus* species growing in rural areas of the UK (Willoughby 2009), at similar climatic conditions in rural Belgium (Ligot et al. 2013) and in urban North America (Dereli et al. 2013), suggest though that annual growth rate increments will vary from c. 20 cm per year for slower growing species such as *Quercus* to 40 cm per year for other faster growing species. These growth rates suggest that urban tree planting schemes that were most frequent in the UK in the 1970s and 1980s (Land Use Consultants 1993; Urban Green Spaces Task Force 2002; Britt and Johnston 2008) could also have contributed to the increased abundance of trees in the 2–5 m and 5–10 m height categories from the 1950s–2010.

The major increase (c. 200%) in the largest trees (>10 m) that occurred from 1900 to 2010 was much less pronounced from 1950 to 2010. This could imply that mortality/removal of larger trees have increased in recent decades, but it also could arise from some variation in the number of larger trees found in 2010 at the locations of the historical photos from the 1900s and 1950s. The occurrence of such stochastic variation is partly driven by the extreme rarity of trees greater than 10 m tall; they account for just 8% of urban trees in 2010. The typical height of mature broad-leaved trees in the UK is much greater than 10 m, for example, ash 20 m, sycamore 24 m, oak 30 m (Fitter and Peat 1994). These three species were the commonest species in Sheffield in 2010 (see Introduction). Our data thus strongly suggest that urban regions are particularly deprived of large old trees, but we still find increases in recent time periods. Moreover, we find a tendency for the largest trees to exhibit greater rates of increase in the areas with most green space, that is, the least urbanized sites. This is presumably because the negative impacts of large urban trees, such as root damage to buildings and street surfaces and the blocking of light, are less likely to occur in the least urbanized sites. It is particularly important to maintain these large trees because of the crucial role they play in providing wildlife resources (Stagoll et al. 2012), cultural ecosystem services (Jim 2004), and their disproportionate contribution to provisioning and regulating services due to their increased biomass (Akbari et al. 2001; Davies et al. 2011).

Space-for-time substitutions (Pickett 1989) are often used in urban ecology to assess the consequences of increasing urbanization intensity over time. The associations we find between rates of increase in tree numbers and urbanization intensity suggest that spatial urbanization gradients may not always provide a reliable measure of change along temporal urbanization gradients. This has important implications for the use of space-for-time swaps in urban systems.

Our data suggest that investment in urban tree planting programmes has contributed to the increase in the number of urban trees over our focal 110-year time period. Maintaining investment in such programmes is thus advisable. This has been achieved in recent years through the Big Tree Plant Campaign which aims to plant an additional one million, mainly urban, trees in England between 2010 and 2015 (http://www.defra.gov.uk/bigtreep lant), but future commitments are uncertain. Moreover, we find some evidence that the smallest trees have increased in abundance the most in areas with little green space, that is, those areas that we also find have the lowest rates of growth in larger trees, which is probably a consequence of increased mortality, for example, tree removal to limit damage to urban infrastructure. Urban tree planting programmes may thus make a larger contribution to future long-term increases in the abundance of old and large trees by giving extra consideration to the potential of planting sites to maintain such trees. Larger trees also contribute disproportionately to ecosystem services, and a more comprehensive and holistic assessment of their benefits may reduce removal rates in situations when tree-associated damage is small relative to the benefits provided by the focal tree. Tree preservation orders in North America have been successful in protecting urban tree stocks when supported by sufficient investment in management and enforcement (Hill et al. 2010; Landry and Pu 2010). In the UK, tree preservation orders can only be applied to trees with high amenity value. This is not precisely defined, but is determined by the suitability of the trees for the focal site, their visibility, and impact, which is a function of factors such as their size, rarity, and screening potential (Department for Communities and Local Government 2006, 2012). Consequently, tree preservation orders are unlikely to be granted for trees in areas with little green space and thus a greater risk of damaging infrastructure or blocking light, or to smaller trees even when surrounded by lots of green space. Enabling preservation orders to be applied to such trees by considering their future rather than just their current amenity value seems likely to reduce tree mortality rates, and further increase the beneficial legacy of urban tree planting programmes by increasing the proportion of such trees that reach full maturity.

REFERENCES

1. Akbari, H., M. Pomerantz, and H. Taha. 2001. Cool surfaces and shade trees to reduce energy use and improve air quality in urban areas. Sol. Energy 70:295–310.
2. Bonnington, C., K. J. Gaston, and K. L. Evans. In press. Squirrels in suburbia: influence of urbanisation on the occurrence and distribution of a common exotic mammal. Urban Ecosyst. doi: 10.1007/s11252-013-0331-2.
3. Britt, C., and M. Johnston. 2008. Trees in Towns II: a new survey of urban trees in England and their condition and management. Department for Communities and Local Government, London.
4. Chen, H., K. Yin, H. Wang, S. Zhong, N. Wu, F. Shi, et al. 2011. Detecting one-hundred-year environmental changes in western China using seven-year repeat photography. PLoS ONE 6:e25008.
5. Dallimer, M., Z. Tang, P. R. Bibby, P. Brindley, K. J. Gaston, and Z. G. Davies. 2011. Temporal changes in greenspace in a highly urbanised region. Biol. Lett. 7:763–766.
6. Davies, R. G., O. Barbosa, R. A. Fuller, J. Tratalos, N. Burke, D. Lewis, et al. 2008. City-wide relationships between green spaces, urban land use and topography. Urban Ecosyst. 11:269–287.
7. Davies, Z. G., J. L. Edmondson, A. Heinemeyer, J. R. Leake, and K. J. Gaston. 2011. Mapping an urban ecosystem service: quantifying above-ground carbon storage at a city-wide scale. J. Appl. Ecol. 48:1125–1134.
8. Department for Communities and Local Government. 2006. Tree Preservation Orders: a guide to the law and good practice. Her Majesty's Stationery Office, London.
9. Department for Communities and Local Government. 2012. Protected trees: a guide to tree preservation procedures. Her Majesty's Stationery Office, London.
10. Dereli, Z., C. Yücedağ, and J. M. Pearce. 2013. Simple and low-cost method of planning for tree growth and lifetime effects on solar photovoltaic systems performance. Sol. Energy 95:300–307.
11. Dodds, J., and D. A. Orwig. 2011. An invasive urban forest pest invades natural environments — Asian long-horned beetle in northeastern US hardwood forests. Can. J. For. Res. 41:1729–1742.
12. Donovan, R. G., H. E. Stewart, S. M. Owen, A. R. Mackenzie, and C. N. Hewitt. 2005. Development and application of an urban tree air quality score for photochemical pollution episodes using the Birmingham, United Kingdom, area as a case study. Environ. Sci. Technol. 39:6730–6738.
13. Evans, K. L. 2010. Urbanisation and individual species. Pp. 53–87 in Gaston. KJ, ed. Urban ecology. Cambridge University Press, Cambridge.
14. Evans, K. L., S. E. Newson, and K. J. Gaston. 2009. Habitat influences on urban avian assemblages. Ibis 151:19–39.
15. Fitter, A. H., and H. J. Peat. 1994. The ecological flora database. J. Ecol. 82:415–425.

16. Forestry Commission. 2001. Inventory report for England. National Inventory of Woodland and Trees. Forestry Commission, Edinburgh.
17. Fuller, R. A., K. N. Irvine, P. Devine-Wright, P. H. Warren, and K. J. Gaston. 2007. Psychological benefits of greenspace increase with biodiversity. Biol. Lett. 3:390–394.
18. Gaston, K. J., M. L. Ávila-Jiménez, and J. L. Edmondson. 2013. Managing urban ecosystems for goods and services. J. Appl. Ecol. 50:830–840.
19. Gillespie, T. W., S. Pincetl, S. Brossard, J. Smith, S. Saatchi, D. Pataki, et al. 2012. A time series of urban forestry in Los Angeles. Urban Ecosyst. 15:233–246.
20. Hall, J. M., J. F. Handley, and A. R. Ennos. 2012. The potential of tree planting to climate-proof high density residential areas in Manchester, UK. Landsc. Urban Plan. 104:140–417.
21. Hill, E., J. H. Dorfman, and E. Kramer. 2010. Evaluating the impact of government land use policies on tree canopy coverage. Land Use Policy 27:407–414.
22. Islam, M. N., K. S. Rahman, M. M. Bahar, M. A. Habib, K. Ando, and N. Hattori. 2012. Pollution attenuation by roadside greenbelt in and around urban areas. Urban For. Urban Greening 11:460–464.
23. Jim, C. Y. 2004. Evaluation of heritage trees for conservation and management in Guangzhou city (China). Environ. Manage. 33:74–86.
24. Kuo, F. E., and W. C. Sullivan. 2001. Environment and crime in the inner city - does vegetation reduce crime? Environ. Behav. 33:343–367.
25. Land Use Consultants. 1993. Trees in towns. HMSO, London, U.K.
26. Landry, S., and R. Pu. 2010. The impact of land development regulation on residential tree cover: an empirical evaluation using high-resolution IKONOS imagery. Landsc. Urban Plan. 94:94–104.
27. Ligot, G., P. Balandier, A. Fayolle, P. Lejeune, and H. Claessens. 2013. Height competition between Quercus petraea and Fagus sylvatica natural regeneration in mixed and uneven-aged stands. For. Ecol. Manage. 304:391–398.
28. Lindberg, F., and C. Grimmond. 2011. Nature of vegetation and building morphology characteristics across a city: influence on shadow patterns and mean radiant temperatures in London. Urban Ecosyst. 14:617–634.
29. Lindenmayer, D. B., W. F. Laurance, and J. F. Franklin. 2012. Global decline in large old trees. Science 338:1305–1306.
30. Maas, J., R. A. Verheij, P. P. Groenewegen, S. de Vries, and P. Spreeuwenberg. 2006. Green space, urbanity, and health: how strong is the relation? J. Epidemiol. Community Health 60:587–592.
31. Meineke, E. K., R. R. Dunn, J. O. Sexton, and S. D. Frank. 2013. Urban warming drives insect pest abundance on street trees. PLoS ONE 8:e59687.
32. Milne, R., and T. A. Brown. 1997. Carbon in the vegetation and soils of Great Britain. J. Environ. Manage. 49:413–433.
33. Monge-Nájera, J., and G. Pérez-Gómez. 2010. Urban vegetation change after a hundred years in a tropical city (San José de Costa Rica). Revista de Biología Tropical 58:1367–1386.

34. Myers-Smith, I. H., D. S. Hik, C. Kennedy, D. Cooley, J. F. Johnstone, A. J. Kenney, et al. 2011. Expansion of canopy-forming willows over the twentieth century on Herschel Island, Yukon Territory, Canada. Ambio 40:610–623.
35. Nowak, D. J. 1993. Historical vegetation change in Oakland and its implications for urban forest management. J. Arboric. 19:313–319.
36. Nowak, D. J., and D. E. Crane. 2002. Carbon storage and sequestration by urban trees in the USA. Environ. Pollut. 116:381–389.
37. Office for National Statistics (2010) Population Estimates for UK, England and Wales, Scotland and Northern Ireland mid-2010 Population Estimates. Available at http://www.ons.gov.uk/ons/rel/pop-estimate/population-estimates-for-uk–england-and-wales–scotland-and-northern-ireland/mid-2010-population-estimates/index.html (accessed 30 September 2011).
38. Oldfield, E. E., R. J. Warren, A. J. Felson, and M. A. Bradford. 2013. FORUM: challenges and future directions in urban afforestation. J. Appl. Ecol. 50:1169–1177.
39. Pauleit, S., R. Ennos, and Y. Golding. 2005. Modeling the environmental impacts of urban land use and land cover change—a study in Merseyside, UK. Landsc. Urban Plan. 71:295–310.
40. Pennisi, E. 2013. Worth a thousand words. Science 341:482.
41. Pickett, S. T. A. (1989) Space-for-time substitution as an alternative to long-term studies. Pp. 110–135 in G. E. Likens, ed. Long-term studies in ecology: approaches and alternatives. Springer-Verlag, New York, USA.
42. Roy, S., J. Byrne, and C. Pickering. 2012. A systematic quantitative review of urban tree benefits, costs, and assessment methods across cities in different climatic zones. Urban For. Urban Greening 11:351–363.
43. Seto, K. C., B. Güneralp, and L. R. Hutyra. 2012. Global forecasts of urban expansion to 2030 and direct impacts on biodiversity and carbon pools. Proc. Natl Acad. Sci. USA 109:16083–16088.
44. Sieghardt, M., E. Mursch-Radlgruber, E. Paoletti, E. Couenberg, A. Dimitrakopoulus, F. Rego, et al. 2005. The abiotic urban environment: impact of urban growing conditions on urban vegetation. Pp. 281–323 in C. C. Konijnendijk, K. Nilsson, T. B. Randrup, J. Schipperijn, eds. Urban forests and trees. Springer, Berlin.
45. Stagoll, K., D. B. Lindenmayer, E. Knight, J. Fischer, and A. D. Manning. 2012. Large trees are keystone structures in urban parks. Conserv. Lett. 5:115–122.
46. Stovin, V. R., A. Jorgensen, and A. Clayden. 2008. Street trees and stormwater management. Arboricultural J. 30:297–310.
47. Ulrich, R. S. 1986. Human responses to vegetation and landscapes. Landsc. Urban Plan. 13:29–44.
48. Urban Green Spaces Task Force. 2002. Green spaces, better places: final Report of the urban green spaces taskforce. DTLR, London.
49. Van Bogaert, R., K. Haneca, J. Hoogesteger, C. Jonasson, M. De Dapper, and T. V. Callaghan. 2011. A century of tree line changes in sub-Arctic Sweden shows local and regional variability and only a minor influence of 20th century climate warming. J. Biogeogr. 38:907–921.

50. Willoughby, I. 2009. The effect of duration of vegetation management on broad-leaved woodland creation by direct seeding. Forestry 82:343–359.
51. Zhou, W., G. Huang, and M. L. Cadenasso. 2011. 90 years of forest cover change in an urbanizing watershed: spatial and temporal dynamics. Landscape Ecol. 26:645–659.

CHAPTER 8

Biological, Social, and Urban Design Factors Affecting Young Street Tree Mortality in New York City

JACQUELINE W.T. LU, ERIKA S. SVENDSEN, LINDSAY K. CAMPBELL, JENNIFER GREENFELD, JESSIE BRADEN, KRISTEN L. KING, AND NANCY FALXA-RAYMOND

8.1 INTRODUCTION

It is understood that the establishment period following planting of an urban street tree is crucial to its survival (Richards 1979; Gilbertson and Bradshaw 1990), yet little is known about the factors or relationships that ultimately contribute to tree mortality or survival. Improving the survival of young street trees can do more to reduce replacement needs than will investments to maintain older trees (Richards 1979). This study of young street trees planted throughout neighborhoods in New York City provides a context in which to understand how biological, social, and urban design factors impact the establishment of new street trees through a multi-disciplinary site assessment framework that examines the conditions of the urban street. In this study, we present our rationale, methods, and descriptive statistics on the subject in an effort to contribute to the literature on street tree health and as a means to inform similar practitioner-based efforts in other urban areas.

One of the fundamental challenges to city managers and civic groups is ensuring the survival of newly-planted street trees in places as dynamic, heterogeneous, and diverse as cities. Population growth, vehicular traffic, poor air quality, and building and sidewalk designs all present challenges to urban street trees, yet trees must reach maturity in order to maximize proven biophysical and social benefits (Dwyer et al. 1992). While there is much research on soil regimes, nursery stock, and species selection, survival rates still vary widely—from 34.7% to 99.7% according to a recent review of the literature (Roman 2006). As cities around the United States increase their investment in tree planting via programs such as MillionTreesNYC, Million Trees Los Angeles, and Keep Indianapolis Beautiful, urban forest managers must be able to ensure young trees' best chance of survival.

Other published work on tree mortality provides insight into factors impacting the life of an urban street tree. One early study analyzes street trees in three Boston neighborhoods that differ both socioeconomically and demographically and reports a 26% mortality rate of 136 trees planted two to four years prior on one commercial street (Foster and Blaine 1978). The authors also observed low rates of vandalism, high rates of automobile damage, and the potential for tree stakes to damage newly-planted trees. Localized effects could also be at play in the findings of an Oakland study that assesses street tree growth and mortality of 480 volunteer-planted trees along a 5.4-mile stretch of one boulevard; after two years, 34% of the trees were dead or removed (Nowak et al. 1990). Although the authors find differences in mortality related to adjacent land uses, it is uncertain if the mortality here is high overall due to conditions local to the boulevard; if the trees were planted incorrectly by the volunteers; or if the trees were too small to withstand minor stresses that may not affect trees of a larger caliper; or some other factor. Another study with a local focus reports on environmental factors influencing 1,000 urban street trees in New York City (Berrang et al. 1985). Because all of the trees in this study are sited directly around electrical power facilities, it is difficult to determine if their observations are a result of this adjacent land use or if they can be applied across the urban landscape. Observational studies such as these give insight into potential factors influencing the survival of newly-planted trees, but have yet to be tested on a city-wide scale. This study examines similarities and differences among a wide range of site conditions and neighborhoods.

The published study with the largest sample size reports on observations of 10,000 newly-planted trees in northern England and finds 9.7% mortality after one year (Gilbertson and Bradshaw 1985). The researchers draw attention to the many factors potentially affecting mortality levels such as stock quality, planting technique, and maintenance regime, but do not attempt to directly link any of these phenomena to tree mortality rates. A similar study tracks four groups of newly-planted trees during their first year in urban Brussels (Impens and Delcarte 1979). The average mortality rate after one year is 11.3%, but detailed information that describes the size, species, or specific location of the trees is not addressed by the study.

A second study about the survival of newly-planted urban trees in Northern England reports on constant, in-situ monitoring of the study trees, which has the potential to provide more detailed information about precisely when and how the tree died (Gilbertson and Bradshaw 1990). The authors found 22.7% mortality after three growing seasons in the inner-city compared with 17% in greater Liverpool. Although the difference is assumed to be linked to the inhospitable environment of the study cohort, vandalism is not a primary cause of tree death in inner city Liverpool. Instead, biological factors such as species tolerance, transplant stress, water stress, and weed competition are deemed most crucial for urban tree establishment (Gilbertson and Bradshaw 1990).

The methods used in urban tree mortality research are broad and varied, making it difficult to compare rates of survival, but several key observations can be gleaned from these prior studies that likely have implications on mortality rates. Vandalism, as measured by the observation of broken branches in the canopy or a broken main stem, is an important factor in the mortality of urban trees (Gilbertson and Bradshaw 1985; Nowak et al. 1990; Pauleit et al. 2002; Roman 2006); adjacent land use can negatively affect street tree populations (Nowak et al. 2004; Roman 2006); and some species of trees fare much better than others as street trees (Gilbertson and Bradshaw 1990; Miller and Miller 1991; Sydnor et al. 1999; Pauleit et al. 2002). Few studies have analyzed the role of physical urban design factors such as traffic volume or the tree's location within the streetscape on mortality rates. Previous studies have not fully investigated the contribution of social or stewardship factors including sociability of the area proximate to the tree (e.g. seating, gardens, front yards) or signs of direct tree care and stewardship (e.g. weeding, mulching, gardening in tree bed), to young street tree success. The goal of this study is twofold, to develop an assessment tool that includes biological, social, and urban design factors and apply it across a wide range of land uses and neighborhood settings to gain insight into the multiple pathways and processes impacting the health of young street trees.

8.2 METHODOLOGY

8.2.1 SAMPLING PLAN

The 13,405 trees analyzed in this study were pulled from a larger sample of 45,094 trees using a partial inventory technique based on stratified random sampling (Sun and Bassuk 1991; Jaenson et al. 1992). The sample was stratified by time in-ground and land use in order to get a random and comprehensive sample of trees in each of these groups. At the time of field survey, all trees had been in the ground between 3 and 9 years. For the stratified random sample, the trees planted from spring 1999 to fall 2003 were grouped into three planting periods. The sample was also stratified using aggregated land use classes from the New York City Primary Land Use Tax Lot Output (PLUTO) data set (NYC Department of City Planning 2005); the original land use types were grouped into One & Two Family Residential, Multifamily Residential; Mixed, Commercial and Public Institutions; Industrial, Utility & Parking; and Open Space & Vacant Land. During field surveys we found that the land use information in PLUTO was not up-to-date or accurate. Forty eight percent of the tree planting locations visited had actual land uses that differed from the PLUTO data. Because of issues encountered with the accuracy of the PLUTO database, we present our results using the land use types observed for the tree in the field. We also readjusted our stratified sample to account for the distribution of field-verified land use.

8.2.2 FIELD METHODS

In order to efficiently visit and record data on 13,405 trees across all five boroughs of New York City, a grid map series at roughly 1:10,000 was produced using ArcGIS. A custom data collection form designed in Pendragon Forms allowed survey questions to be loaded on a Palm Pilot for mobile data collection. These field data were directly synchronized into Microsoft Excel. In this study, the data were collected at multiple scales—the tree level, then the building level, and at the block level. In order to facilitate easy repetition of data collection, all variables were optimized for simple field observation and require no laboratory analysis or precise measurements. The data are organized into the three groups of relevant information: biological factors that may affect young street trees, urban design factors, and sociability/stewardship factors. Some of the variables we collected can apply to more than one tier—for example, presence or absence of a tree guard can be both a physical design and a stewardship factor, depending on whether they are routinely installed as part of municipal tree planting.

These methods were based upon social site assessment models used for natural resource management (Freudenburg 1986) with city foresters taking an active role in training and supervising researchers in the field. All fieldwork was conducted by 20 interns hired and trained by the New York City Department of Parks & Recreation (NYC Parks) and the USDA Forest Service Northern Research Station (NRS). Data collection took place over the summers of 2006 and 2007 in hundreds of New York City neighborhoods. Recording the presence or absence of observable phenomena, the team used a combined study approach and developed a data collection framework that resulted in the collection of over forty items of data at the location of each tree. Street tree locations varied widely, from high-rise areas, to low-rise brownstone neighborhoods, to single family structures in suburban settings. For the purposes of this analysis, missing trees were counted as dead, following the precedent of previous studies (Gilbertson and Bradshaw 1990; Miller and Miller 1991; Pauleit et al. 2002).

8.2.3 BIOLOGICAL FACTORS

Table 1 lists the biological factors that may have an effect on the success and failure of young street trees. If the tree cannot obtain its minimum biological requirements, it will not thrive, regardless of the urban context in which it was planted. This first layer of data collection provides important clues to the overall health of the tree. The data items listed below may indicate tree health, growth rates, damage and decay, or soil health or identify biological stressors affecting establishment. They are most useful in determining the overall health of a living street tree; if a tree is dead or missing from where it was planted, it is not possible to collect many of these data items. In light of the developing awareness in an objective methodology in appraising tree health (Bond 2010) and linking urban tree evaluations into the forest inventory analysis (FIA) through the ongoing International Union of Forest Research Organizations (IUFRO) Urban Forestry Data Standards effort, our approach is certainly subject to change as methods become standardized. Soil compaction was measured by applying pressure to the soil with a screwdriver tip; if the screwdriver easily entered the soil, the soil was said to be uncompacted.

8.2.4 SOCIABILITY/STEWARDSHIP FACTORS

The social factors which potentially influence young street tree mortality are listed in Table 2. Our data collection methodology includes recording direct signs of tree stewardship at the level of each tree (i.e. planting in tree pits, adding mulch), which are indicators that individuals or groups are caring for a tree. At the building and neighborhood level, we observed off-tree signs of stewardship such as the presence of home decorations, front yard gardens, and murals. These factors are considered—"cues to care" that provide evidence that individual and/or community-level stewardship is taking place (Nassauer 1995). A well-cared for urban street tree and pit area is considered to be a sign of active local stewardship. We also collected data on practices that could have conflicting effects on a tree's health; for example, tree lights could retard tree growth by strangling the

Table 1. Biological factors potentially affecting young street trees in NYC.

Data Item	Response type
water pooling in tree pit	presence/absence
soil compaction	presence/absence
animal waste	presence/absence
sucker growth	presence/absence
evidence of leaf chlorosis	presence/absence
evidence of insect damage	presence/absence
evidence of dieback	presence/absence
guiding wires girdling tree	presence/absence
guard/grate girdling tree	presence/absence
broken branches	presence/absence
unnatural lean	presence/absence
trunk wound	presence/absence
pit soil level	categorical
planting depth	categorical
species	categorical
diameter at breast height	categorical

tree, but also could draw attention to the presence of a tree thereby triggering stewardship.

Data were collected about neighborhood sociability to ascertain whether the tree is incorporated into active street life. For example, benches are built into tree pits, seating is arranged under trees' canopies, or play equipment is often proximate to the tree. At the neighborhood level, signs of sociability indicate more—"eyes upon the street" (Jacobs 1961) or the orientation of urban space to enhance community awareness and engagement. This sociability can influence tree survival via multiple pathways, such as through prevention of tree vandalism. Moreover, these signs of sociability can be considered indicators of community street life and may relate to stewardship over time. Given a study that collects observational data at one moment in time, it is important to use these proximate measures of social life as indicators

Table 2. Sociability/stewardship factors potentially affecting street trees in NYC.

	Data Item	Response type
Tree/Tree pit level	pit off curb (at least 12" away)	presence/absence
	curb intact	presence/absence
	tree grate	presence/absence
	block paving in tree pit	presence/absence
	tree guard*	presence/absence
	tree pit type	categorical
	presence/condition of block pavers	presence/absence; categorical
	tree pit size (square feet)	number
Building level	ground floor door	presence/absence
	awning on adjacent building	presence/absence
	scaffolding on adjacent building	presence/absence
	number of building stories	number
	land use classification	categorical
Streetscape level	median strip on street	presence/absence
	on-street parking	presence/absence
	bus stop nearby (< 5')	presence/absence
	driveway nearby (< 5')	presence/absence
	bike rack nearby	presence/absence
	sidewalk condition	categorical
	traffic volume	categorical
	tree placement in slope	categorical
	sidewalk width	number
	number of traffic lanes	number
	% pavement within drip line	number

* the variable presence of a tree guard can also apply to the sociability/stewardship category.

that stewardship may have occurred historically. Areas of community street activity include facilities such as places of worship and schools, which are known to sponsor local stewardship activities. Drawing upon the work of Wilson and Kelling (1982), negative indicators were also observed, such as the presence of broken windows, vacant lots and buildings, and (non-mural) graffiti. Known as the—"broken-window theory," the presence of vacant buildings and lots strewn with garbage tend to attract more visible disorder on and around neighborhood streets. Researchers documented the presence and absence of disorder around each street tree.

One difference in this section of data is that it is possible for some items to have two response types. For example, if a front yard is present (presence/absence), it may be valuable to note what type of yard (categorical; i.e. paved, grass). The same can be said for gardens, building security, murals, and public facilities. Collecting this second tier of data gives researchers the ability to strengthen an analysis of the dynamic social factors affecting street tree mortality.

8.2.5 URBAN DESIGN FACTORS

This study suggests that physical urban design factors influence the success of young street trees; this category includes information at three different levels: tree/tree pit, building, and streetscape (listed in Table 3). The factors measured at the level of the tree and tree pit itself are more directly connected with the tree success or failure, while others, such as the presence of a bike rack nearby and the width of the sidewalk, are more exploratory in nature and may only provide insights into potential influences. All factors comprise the physical urban context into which the tree has been planted. They are the result of urban design, zoning practices, or unplanned piecemeal development and they affect the flow of pedestrians, bicycles, and motor vehicles through the environment surrounding the tree. At the same time, these factors also affect airflow, sunlight, and wind speed that can impact the growing conditions of trees (McGrath et al. 2007).

Most of these data are collected in the presence/absence format, but some other responses are categorical in nature. For example, pit type could be characterized as a sidewalk cutout or tree lawn; block paving status can range from good to raised or altogether missing; traffic volume could

Table 3. Urban design factors potentially affecting street trees in NYC.

	Data Item	Response type
Tree/Tree pit level	tree care-related signage	presence/absence
	stakes present, but no wires	presence/absence
	walled tree well	presence/absence
	tree pit plantings	presence/absence
	tree guard*	presence/absence
	tree pit paved to tree trunk	presence/absence
	mulch in tree pit	presence/absence
	gravel in tree pit	presence/absence
	bench near/around pit	presence/absence
	bird feeder in tree or tree pit	presence/absence
	irrigation bag	presence/absence
	evidence of weeding of tree pit	presence/absence
	litter in tree pit	presence/absence
	evidence of pruning	presence/absence
	debris in canopy of tree	presence/absence
	electrical outlet in tree pit	presence/absence
	lights in or around tree	presence/absence
Building level	seating area associated with building	presence/absence
	play equipment in yard of building	presence/absence
	flag on building	presence/absence
	decorations on door of building	presence/absence
	flower planters	presence/absence
	building has front yard (type)	presence/absence; categorical
	building has garden (type)	presence/absence; categorical
	building security (type)	presence/absence; categorical
Streetscape	graffiti on adjacent buildings	presence/absence
	broken/missing windows	presence/absence
	mural on adjacent building (type)	presence/absence; categorical
	public facilities on block (type)	presence/absence; categorical
	block-level vacancies	categorical

* the variable presence of a tree guard can also apply to the urban design category.

be low, medium, or high; and sidewalk condition could be good, cracked, poor condition, etc.

8.3 FINDINGS FROM DESCRIPTIVE STATISTICS

The following descriptive statistical analyses examine the effects of time since planting, land use, and selected biological, social, and urban design factors on urban young street tree mortality. Contingency tables and chi-square analyses were used to assess the effect of each variable, with the simplifying assumption that variables are independent and do not interact with each other. Although in reality our dataset contains many nested, correlated and confounding variables, as practitioners we are interested in evaluating the contributions of each variable from a management perspective and for refining planting policies and site selection procedures. Formal analysis incorporating combinations of and interactions between these factors is ongoing and will be treated in future manuscripts.

8.3.1 TIME SINCE PLANTING

As previously mentioned, it is widely assumed in the literature that there is some time after planting in which the mortality rates of street tree populations stabilize. In order to determine if and possibly when this is occurring in New York City, we performed a preliminary analysis to determine if time since planting is related to street tree mortality. Our data do in fact suggest this type of trend, as the rate of tree loss for trees inspected 6-8 and 8-9 years after planting are nearly identical. Contingency table analysis found years since planting to have a significant influence on tree survival (Pearson's $X^2=24.65$, $df=2$, $p<0.001$). The decrease in survival rate between the first two time periods is the most marked, which reflects the immediate difficulty that young street trees face after being transplanted into the urban landscape. The two-year survival rate for these young street trees was calculated using operational contract data.

Table 4. Young street tree survival by years since planting.

Years since planting	Alive		Not Alive		Total sample size
	No. of trees	%	No. of trees	%	
2 years after planting*	41,169	91.3%	3,925	8.7%	45,094
3-6 years after planting	1,891	78.2%	526	21.8%	2,417
6-8 years after planting	3,690	73.0%	1,363	27.0%	5,053
8-9 years after planting	4,381	73.8%	1,554	26.2%	5,935
Total	9,962	74.3%	3,443	25.7%	13,405

* 2 year survival rate is based on contractual guarantee inspection data and is only provided for reference.

8.3.2 LAND USE

Because previous research highlighted the importance of adjacent land use in young street tree mortality, we performed an additional analysis examining this phenomenon in New York City. For this analysis, observed land uses were grouped into five categories: one/two family residential; multi-family residential; mixed, commercial, and public institutions; industrial, utility, and parking; and open space/vacant land.

In New York City, young street trees in one and two family residential areas have the highest survival rate (Table 5), while industrial areas and open space/vacant land had the lowest rates of street tree survival (ranging from 60.3% to 62.9%). Pearson's chi-square test found land use group to have a significant influence on tree survival (X^2=455.432, df=4,p<0.001). This data suggests that neighboring human activities do have an effect on young street tree survival and our results are similar to those found in other studies (e.g. Nowak et al. 1990; Nowak et al. 2004).

Table 5. Young street tree survival by years since planting.

Land Use Group	Alive		Not Alive		Total sample size
	No. of trees	%	No. of trees	%	
One/Two Family Residential	4,821	82.7%	1,009	17.3%	5,830
Multi-Family Residential	2,232	72.3%	856	27.7%	3,088
Mixed, Commercial and Public Institutions	388	62.9%	229	37.1%	617
Industrial, Utility and Parking	1,903	66.2%	972	33.8%	2,875
Open Space and Vacant Land	545	60.3%	359	39.7%	904
Total	9,889	74.3%	3,425	25.7%	13,314

* 2 year survival rate is based on contractual guarantee inspection data and is only provided for reference.

8.3.3 BIOLOGICAL, SOCIABILITY/STEWARDSHIP, AND URBAN DESIGN FACTORS

As mentioned previously, we looked at how individual or groups of variables affected survival rates through a series of two-way contingency tables. The results presented here begin to lay out the type of processes at work in the urban forest. Our initial results are summarized in Tables 6 through 8.

8.3.3.1 BIOLOGICAL FACTORS

Previous research has shown that species does matter with respect to the mortality of urban street trees, and this study reinforces that idea that there are significant differences in survival rates between species (Table 6). Of

Table 6. Young street tree survival by years since planting.

Independent Variable	Alive	Not Alive	% Survival	X^2 value	df	p -value
Tree species (>1% of all planted trees)						
Pyrus calleryana	1,863	381	83.0%			
Gleditsia triacanthos	1,274	332	79.3%			
Tilia cordata	617	168	78.6%			
Quercus palustris	639	177	78.3%			
Zelkova serrata	537	149	78.3%			
Tilia tomentosa	143	41	77.7%			
Quercus rubra	145	42	77.5%			
Fraxinus pennsylvanica	268	85	75.9%	178.611	18	<0.001
Prunus cerasifera (Purpleleaf plum)	113	37	75.3%			
Acer rubrum	245	81	75.2%			
Prunus serotina (Kwanzan cherry)	266	88	75.1%			
Japanese pagoda tree	310	109	74.0%			
Prunus virginiana (Shubert cherry)	452	184	71.1%			
Tilia tomentosa	477	204	70.0%			
Acer campestre	170	73	70.0%			
Liquidambar styraciflua	171	77	69.0%			
Prunus spp.	210	107	66.2%			
Gingko biloba	370	189	66.2%			
Plantanus acerifolia	112	68	62.2%			
Presence of animal scat in tree pit or near tree						
Present	627	139	81.9%	24.19	1	<0.001
Not present	9,335	3,301	73.9%			

the trees planted that comprise greater than one percent of the total, callery pear (*Pyrus calleryana*) is the most successful. Although the entire suite of species that NYC Parks plants are known to be tolerant of urban conditions, some have higher tolerances than others. Anecdotally, one of the most common stressors that an urban street tree faces believed to face is deposition of animal waste in the tree pit, yet in our results the presence of scat was unexpectedly associated with higher survival, underscoring how these simplistic analyses based on one-time observations should be interpreted with caution.

8.3.3.2 SOCIABILITY/ STEWARDSHIP FACTORS

These variables can help to elucidate the level of engagement that an individual or local community group has with trees in the urban landscape. In terms of sociability, trees with adjacent seating or an adjacent front yard were all more likely to survive in the urban environment (Table 7). Our data also show that a tree is more likely to survive if the building in front of which it is planted has a garden or planters/window boxes. If a garden is present, though, the type or visible level of garden care does not have any bearing on young street tree survival. Our interpretation of these results is that either (1) the mere presence of adjacent stewardship of other natural amenities (lawns, gardens) is adequate to engage local residents in the care of maintenance of their street trees; or (2) presence of signs of off-tree stewardship may be an indicator of on-tree stewardship that has occurred historically.

A stewardship index was constructed from factors that directly affect the area in and around the tree pit, including: presence of signage, plantings in pits, mulch, and evidence of weeding. This stewardship index is significantly correlated with tree survival. Planting in the tree pit was the most often observed stewardship behavior (1,039 trees), followed by mulch (962 trees), weeding (317 trees), and signage (232 trees). Evidence of active, direct tree stewardship is a positive indicator or predictor of street tree survival.

8.3.3.3 URBAN DESIGN FACTORS

Our research indicates that the urban context into which street trees are planted is an important factor in their success and failure (Table 8). Street trees have a greater chance at survival when planted in lawn strips rather than sidewalk cutouts. In our data the size of sidewalk cut out pits does not have a significant influence on the survival of young street trees. Given

Table 7. Young street tree survival and select sociability/stewardship factors

Independent Variable	Alive	Not Alive	% Survival	X^2 value	df	p -value
Presence of seating near tree						
With seating	694	135	83.7%	28.44	1	<0.001
No seating	8,719	2,824	75.5%			
Presence of front yard near tree						
Yard present	5,246	1,170	81.8%	236.40	1	<0.001
No yard	4,167	1,789	70.0%			
Presence of a garden near tree						
Garden present	3,266	607	84.3%	210.59	1	<0.001
No garden	6,147	2,352	72.3%			
Garden type (if present)						
Natural	3,345	623	84.3%	1.04	1	0.308
Plastic	12	4	75.0%			
Garden care (if present)						
Good	3,201	580	84.7%	4.40	1	0.036
Poor	155	41	79.1%			
Presence of planters or window boxes						
Present	1,623	244	86.9%	142.19	1	<0.001
Not present	7,790	2,715	74.2%			
Presence of stewardship signs*						
4 signs	20	0	100.0%			
3 signs	112	3	97.4%			
2 signs	328	11	96.8%	412.36	4	<0.001
1 sign	1,325	122	91.6%			
None	8,177	3,307	71.2%			

* signs of stewardship include presence of signage on or near the tree; plantings in street tree pits; mulch placed in pit; and evidence of weeding.

that larger tree pits yield greater volumes of uncompacted soil for the roots to grow and greater surface area for water to enter the tree pit, one would expect that street trees would fare much better in large tree pits. One possible interpretation of this result is that tree pit size is not as important in the early life of a young street tree, but will become a limiting factor as the tree begins to grow out of its spot in the sidewalk.

Table 8. Young street tree survival and select sociability/stewardship factors

Independent Variable	Alive	Not Alive	% Survival	X^2 value	df	p -value
Pit type						
Lawn	3,548	992	78.1%			
Sidewalk	5,917	2,196	72.9%	58.43	2	<0.001
Continuous	397	193	67.3%			
Presence of perimeter tree guard						
With guard	1,121	83	93.1%	116.42	1	<0.001
No guard	8,841	2,150	80.4%			
Tree Pit Size (sidewalk trees only)						
55+ sq. ft	42	7	85.7%			
45 to <55 sq. ft	160	29	84.7%			
15 to <25 sq. ft	3,066	570	84.3%			
05 to <15 sq. ft	336	70	82.8%	7.48	5	0.188
35 to <45 sq. ft	266	58	82.1%			
25 to <35 sq. ft	2,007	446	81.8%			
Tree location						
Located on curb	9,413	2,959	76.1%	262.78	1	<0.001
Located on median	549	484	53.1%			
Observed traffic volume						
Light	6,785	1,842	78.6%			
Moderate	2,224	1,026	68.4%	280.49	2	<0.001
Heavy	806	530	60.3%			

Installing a perimeter tree pit guard prevents vandalism and vehicular damage, prevents animal waste deposition, and is visually representative of a tree that is being cared for by someone. It is likely because of a combination these factors that trees in pits with perimeter guards have a greater chance at success than trees in unprotected pits. The presence/absence of tree guards can also be considered as a sociability/stewardship factor, not just a physical design variable. This is because while the mechanism for reduced mortality for street trees with tree guards are physical (by preventing soil compaction or inadvertent contact to the tree by cars), tree guards are typically installed privately and not by NYC Parks, and therefore also represents an act of stewardship. This may vary in other urban areas.

The physical location of the tree within the urban streetscape is also significant. Trees planted in street medians have a poor chance at survival when compared to trees planted at the curbside. Traffic volume also has an effect on young street tree mortality, with trees in low traffic areas faring better than those planted in moderate or high traffic thoroughfares.

Another finding not explored here but worthy of discussion is that of missing trees. Of the over 13,000 trees visited in this study, nearly twenty percent of them were not present from their planted location while only six percent were standing dead. Although these two groups were collapsed for the purpose of discussing overall mortality, their large number warranted further analysis. We looked at whether or not the populations of standing and dead trees were significantly different with respect to some of our variables and found the following: trash in the tree pit is more common with dead trees than missing; missing trees are more likely when a sidewalk is less than five feet wide; trees are more likely to be missing than standing dead in a lawn strip than any other pit type. Missing trees are not statistically linked to the following: street slope, presence of street parking, sidewalk condition, or traffic volume. Urban forest managers in New York City agree that there are several possibilities of the fate of those missing trees: vandalism, vehicular collision, or tree removal without subsequent replacement but, regardless of the pathway, these missing trees are dead.

8.4 DISCUSSION

The highly local and specific nature of other published street tree mortality studies inspired this study to examine which factors may affect mortality in New York City. New York City's street tree planting mortality rates are lower than those published for other cities (see Figure 1). Some possible reasons for this distinction are: trees planted in New York City are planted by experienced contractors working under the supervision of trained foresters, while other tree planting programs frequently use volunteers with little or no planting experience (e.g. Nowak et al. 1990) or aren't working with strict contract specifications; and larger caliper trees (2.5-3") are planted in New York City, while smaller stock was planted in other locations (Nowak et al. 1990; Gilbertson and Bradshaw 1990).

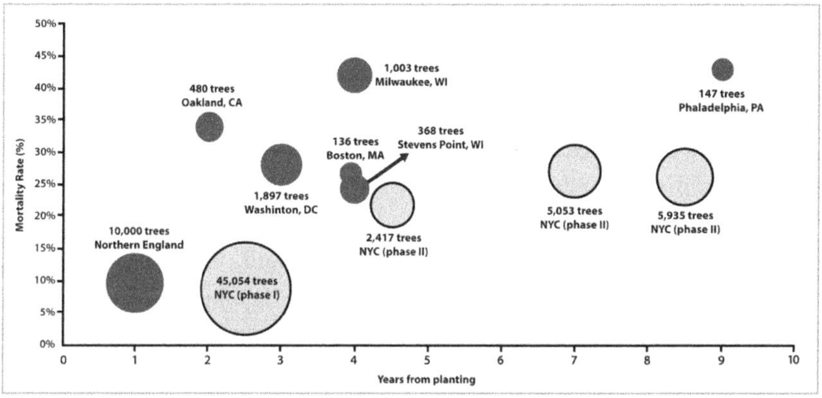

FIGURE 1: Other newly-planted street tree mortality studies (Aggregated from Roman, 2006), including the results from New York City.

In this manuscript we present a socio-ecological-design framework for future young street tree mortality research, with the intention of facilitating the replication of this type of study in other urban areas. Based on this work we have developed a Site Assessment Tools Description (available at http://www.nyc.gov/parks/ystm), a step-by-step guide for city managers and researchers on how to assess early street tree survival and mortality. Our hope is that other cities will replicate at least part of this study and over time build up data sets which will allow for cross-city comparisons.

These preliminary results provide an initial understanding of some of the factors that are important in the success and failure of young street trees planted in New York City, and provides direct feedback that managers can use to refine NYC Parks' planting practices and policies. Variation in planting survival rates by species has important implications for the long-term dynamics of New York City's street tree population. In terms of a tree's urban design and neighborhood context, this study confirms the observations of many urban foresters that curbside trees planted in lawn strips and in low-vehicular traffic areas are more likely to survive. This study also quantifies the disproportionately high mortality rates of trees that are planted in street medians compared to trees located on the curb. Based on this result, NYC Parks has already changed their planting policies for median trees, and is planting trees in only the widest street medians, where adverse factors like collisions, salt exposure, and minimal soil volume are less likely. Similarly, our observation of the effectiveness of tree guards in protecting young street trees is corroborated by the experiences of NYC's practicing urban foresters. Such demonstrated effectiveness may justify the expense of securing street tree guards at the time of planting.

Our results suggest that civic stewardship and neighborhood sociability is a critical complement to municipal management and investment in new street tree plantings. However, we have only started to explore how the data we collected could be used to develop more comprehensive indices representing stewardship or neighborhood sociability. The mechanisms that relate the signs of neighborhood sociability—or even of other non-tree signs of stewardship—to improved tree survival cannot be revealed through this study. While we hypothesize that active presence of residents on the street can serve to help ensure that vandalism of trees does not occur, other qualitative methods such as interviews and repeated social observational studies would be required to evaluate this hypothesis. Moreover, this study cannot determine directionality of observed relationships. For example, the presence of stewardship activities in nearby lawns and gardens may either inspire the care of street trees, or the presence of the new tree itself may encourage other acts of local stewardship along the street.

The initial results presented here offer an important basis for urban planning programs as well as for researchers interested in further exploring factors affecting tree canopy restoration efforts in the urban environment. This is just the beginning of what we will be able to learn from the data we collected using this integrated socio-ecological framework. The current MillionTreesNYC campaign aims to plant street trees in every available and feasible sidewalk location across a wide range of site types in New York City, but at other times and in other places, difficult choices must be made in terms of street tree planting locations. Taken together, these biological, social, and urban design factors can be weighed by urban foresters when designing and selecting the locations for street tree plantings and developing community stewardship programs. Further analysis of our data set will assess the relative importance of these and the remaining data variables that were collected during the field survey of these trees. As cities such as New York continue to develop and implement comprehensive tree planting campaigns, these findings provide insight in the field of natural resource management on the relationship between locations and vulnerability; stewardship and sustainability.

REFERENCES

1. Berrang, P., D.F. Karnosky, and B.J. Stanton. 1985. Environmental factors affecting tree health in New York City. Journal of Arboriculture 11(6):185-189.
2. Dwyer J.F., E.G. McPherson, H.W. Schroeder, and R.A. Rowntree. 1992. Assessing the benefits and costs of the urban forest. Journal of Arboriculture. 18(5):227-234.
3. Fisher, D.R., E.S. Svendsen, and L.K. Campbell. 2007. Toward a framework for mapping urban environmental stewardship. International Symposium on Society and Resource Management, published abstract. http://www.docstoc.com/docs/50284639/List-of-Oral-Presentations-ISSRM-2007-(listed-in-alphabetical (accessed 01/28/2011).
4. Foster, R.S. and J. Blaine. 1978. Urban tree survival: trees in the sidewalk. Journal of Arboriculture 4(1): 14-17.
5. Freudenburg, W.R., 1986. Social impact assessment. Annual Review of Sociology 12:451-478.
6. Gilbertson, P. and A.D. Bradshaw. 1985. Tree survival in cities: the extent and nature of the problem. Arboricultural Journal 9:131-142.
7. Gilbertson, P. and A.D. Bradshaw. 1990. The survival of newly-planted trees in inner cities. Arboricultural Journal 14:287-309.
8. Impens, R.A. and E. Delcarte. 1979. Survey of urban trees in Brussels, Belgium. Journal of Arboriculture 5(8):169-176.
9. Jacobs, J. 1961. The Death and Life of Great American Cities. 1992 edition, Vintage Books, New York.
10. Jaenson, R., N. Bassuk, S. Schwager, and D. Headley. 1992. A statistical method for the accurate and rapid sampling of urban street tree populations. Journal of Arboriculture 18(4):171-183.
11. McGrath, B., V. Marshall, M.L. Cadenasso, J.M. Grove, S.T.A. Pickett, R. Plunz, and J. Towers. 2007. Designing Patch Dynamics. Columbia University, School of Architecture, Planning and Preservation. New York. 250 pp.
12. Miller, R.H. and R.W. Miller. 1991. Planting survival of selected street tree taxa. Journal of Arboriculture 17(7):185-191.
13. Nassauer, J.I. 1995. Messy ecosystems, orderly frames. Landscape Journal 14(2):161-169.
14. Nowak, D.J., J.R. McBride, and R.A. Beatty. 1990. Newly planted street tree growth and mortality. Journal of Arboriculture 16 (5):124-129.

15. Nowak, D.J., M. Kuroda, and D.E. Crane. 2004. Tree mortality rates and tree population projections in Baltimore, Maryland, USA. Urban Forestry and Urban Greening 2:139-147.
16. NYC Department of City Planning. 2005. MapPLUTO (Release 04C): NYC Department of City Planning.
17. Pauleit, S., N. Jones, G. Garcia-Martin, J.L. Garcia-Valdecantos, L.M. Rivière, L. Vidal-Beaudet, M. Bodson, and T.B. Randrup. 2002. Tree establishment practice in cities and towns -- Results from a European survey. Urban Forestry and Urban Greening 5(3):111-120.
18. Richards, N.A. 1979. Modeling survival and consequent replacement needs in a street tree population. Journal of Arboriculture 5(11):251-255.
19. Roman, L. 2006. Trends in street tree survival: Philadelphia, PA. University of Pennsylvania, Department of Earth and Environmental Science, Master of Environmental Studies Capstone Project. 30 pp. http://repository.upenn.edu/mes_capstones/4/. (accessed 01/28/2011).
20. Sydnor, D., J. Chatfield, D. Todd, and D. Balser,. 1999. Ohio street tree evaluation project. Ohio State University and Ohio Department of Natural Resources. Bulletin 877-99. http://www.dnr.state.oh.us/forestry/urban/ostep/ostepintro/tabid/5546/Default.aspx. (accessed 05/01/2010).
21. Sun, W.-Q. and N. Bassuk. 1991. Approach to determine effective sampling size for urban street tree survey. Landscape and Urban Planning 20(4):277-283.
22. Wilson, J.Q. and G. Kelling. 1982. The police and neighborhood safety: Broken windows. Atlantic Monthly 127:29-38.

CHAPTER 9

Tree Diversity in Southern California's Urban Forest: The Interacting Roles of Social and Environmental Variables

MEGHAN L. AVOLIO, DIANE E. PATAKI, THOMAS W. GILLESPIE, G. DARREL JENERETTE, HEATHER R. MCCARTHY, STEPHANIE PINCETL, AND LORRAINE WELLER CLARKE

9.1 INTRODUCTION

Urban forests are unique in that they are novel assemblages of native and exotic tree species (Kunick, 1987; Jim, 1993; Sjöman et al., 2012; Aronson et al., 2015) that are influenced by both biophysical (e.g., climatic factors) and human drivers (e.g., management and planting preferences; Sanders, 1984; Kunick, 1987; Talarchek, 1990). Accordingly, both socio-economic and environmental drivers are necessary to explain patterns of urban forest composition and cover. Within different cities, studies have found a negative relationship between tree cover and population density (Iverson and Cook, 2000; Clarke et al., 2013), a positive relationship between tree cover and income (Talarchek, 1990; Iverson and Cook, 2000; Lowry et al., 2011; Clarke et al., 2013), a positive relationship between tree cover and home or neighborhood age (Lowry et al., 2011), and a postivie relationship between tree cover and education (Heynen and Lindsey, 2003; Luck et al., 2009; Kendal et al., 2012b). To our knowledge, fewer studies have found relationship with urban forests and environmental drivers. In Salt Lake Valley, UT, Lowry et al. (2011) found greater tree cover in areas of higher precipitation, while in Los Angeles, CA, Clarke et al. (2013) found no relationship between tree cover and distance from the coast, an integrative measure of environmental conditions. Heynen and Lindsey (2003) found greater tree cover in areas with higher stream density and steeper slopes across urban areas in central Indiana. In addition to overall tree cover, understanding how sociological and biophysical drivers affect species richness and measures of community diversity is necessary for understanding the composition and drivers of urban forests.

Controls of diversity have been investigated in natural systems worldwide (Gaston, 2000), and the factors that influence diversity can vary depending on the scale being investigated (Whittaker et al., 2001; Field et al., 2009). Many drivers have analogs in urban ecosystems. Area is a key determinant of diversity, where larger areas can support more species (Gaston, 2000; Whittaker et al., 2001). In cities, population density can be indicative of available area for vegetation, as the aerial extent of vegetation generally declines with population density within cities (Jenerette et al., 2007). Time since disturbance is an important determinant of diversity (Whittaker et al., 2001), in that species richness increases during primary succession (Anderson, 2007). In cities, species richness has been shown to increase with time since development (Martin et al., 2004; Boone et al., 2009; Kirkpatrick et al., 2011; Clarke et al., 2013). This may be seen as analogous to time since disturbance, and therefore there are parallels between processes of community assembly in more natural vs. urban ecosystems. Household income has also been shown to positively correlate with species richness (Hope et al., 2003; Martin et al., 2004; Cook et al., 2012) and this relationship has been termed the "luxury effect" (Hope et al., 2003). However, how this relationship might be integrated into existing theories about ecological determinants of diversity in natural systems is less clear. Although, relationships between education and tree cover have been found, to our knowledge no study has linked education to biodiversity. How these different drivers of diversity influence tree diversity at different scales, from plot to neighborhood, municipal and larger regional scales, has not been investigated.

The urban forest as a whole can be considered a mosaic of smaller land parcels or patches, either public or private, that each have their own unique set of drivers (Sanders, 1984; Zipperer et al., 1997). On a large scale, municipalities or districts within cities have different levels of tree diversity (Jim and Liu, 2001a; Bourne and Conway, 2013). Within these governmental designations, tree diversity, richness, and species identity can also differ among land use types (e.g., commercial vs. transportation; Bourne and Conway, 2013; Clarke et al., 2013) and private (e.g., residential) vs. public trees (e.g., street trees) that are typically managed by the municipality (Maco and McPherson, 2002). Both residential and street trees have their own unique set of drivers (Roman and Scatena, 2011; Pincetl et al., 2012), which is reflected in street and residential trees having different traits and species composition (Jim, 1993; Kirkpatrick et al., 2011). Differences between street and residential trees likely reflect the different planting pressures and preferences of the city and private land owners (Jim, 1993; Kirkpatrick et al., 2011). Residential and street trees are chosen based on management requirements and desires for specific attributes (McBride and Jacobs, 1976), which change over time (McBride and Jacobs, 1976; Kunick, 1987; Pearce, 2013). Thus, there are multiple spatial scales, from municipalities to residential parcels, at which human preferences shape tree communities.

As ecosystems are transformed to urban areas, a series of filters change the composition of the component plant species, both limiting which species can survive in the new urban environment and adding desirable species that are planted (Williams et al., 2009). In arid and semi-arid cities, trees are not a key component of the native ecosystem and urban trees are planted as the city is built. In comparison, in mesic areas where trees are native, trees were removed as the city is built. Thus, key to predicting patterns of urban biodiversity is to understand resident preferences and how these preferences are reflected in plant assemblages, especially in arid and semi-arid cities. While studies have noted that the high proportion of flowering or fruit trees reflect resident desires for these traits (Jim, 1993; Cook et al., 2012), no study, to our knowledge, has directly linked resident preferences with patterns of urban tree diversity. In Australia, Kendal et al.

(2012a) found that homeowner garden preferences were correlated with garden plant traits, and these relationships were stronger for residents who owned their homes for longer periods of time. It is unknown, however, if there are similar patterns across larger regional scales and for other types of planted urban vegetation.

Here, we investigated patterns of urban tree community composition in the Los Angeles Metropolitan area of southern California to determine whether these patterns are correlated with resident preferences. In a previous study, over 1000 residents across southern California were surveyed about their preferences for tree attributes (Pataki et al., 2013; Avolio et al., 2015). Low-income residents had a higher preference for fruit trees, and higher income residents expressed greater preferences to have trees in their yard than low-income residents (Avolio et al., 2015). Additionally, preferences for tree attributes were affected by local climatic conditions. For example, residents who lived in hotter areas had a greater preference for shade trees, and residents who lived in drier areas had a greater preference for trees that used less water (Avolio et al., 2015). Overall, certain tree attributes were more important than others, with aesthetics and provision of shade ranked particularly highly (Pataki et al., 2013; Avolio et al., 2015), however, we do not know whether resident preferences for tree attributes are actually reflected in compositional patterns of urban forests.

Traits are increasingly used in ecological studies (McGill et al., 2006) to understand plant distributions and responses to environmental change (Reich et al., 1997; Díaz et al., 1998) and urbanization (Vallet et al., 2010). Traits commonly used in ecological studies (Cornelissen et al., 2003), however, are not necessarily best suited for urban research, as many of these traits have no direct correlate with attributes chosen by city managers or residents (e.g., aesthetic attributes). To overcome this potential limitation, Pataki et al. (2013) proposed "ecosystem service-based traits" that are linked to known resident preferences, including: water requirements, size at maturity, and presence of showy flowers. Similarly, Zhang and Jim (2014), used similar traits and called them "ecological amenities," evaluating whether urban trees provided seasonal changes in foliage color, shading, and edible fruit. In Taipei, Jim and Chen (2008) found that the main function of certain tree communities was to beautify the surrounding area. In Guangzhou, China, Jim and Liu (2001b) found that the most important ecological amenity in roadsides was shade, while in parks it was flower or fruit provision, demonstrating that different areas are managed to provide different functions. Thus, studying tree traits that are important for land mangers (e.g., city parks departments, residential home owners) may be key to a predictive understanding of community composition patterns of urban forests.

We had three objectives in this study. Our first objective was to investigate the relative importance of socio-economic and environmental factors in determining patterns of community diversity and cover of the urban forest in southern California at county and regional scales. Our second objective was to determine whether there are discernable spatial patterns in urban forest species composition and the distribution of traits. We hypothesized that there would be differences among counties and tree types (street vs. residential trees) in species composition and functional trait richness because of different actors involved in making planting decisions. We also hypothesized that traits would be more useful for differentiating between counties and tree types than species composition because the majority of trees in LA are planted rather than naturally regenerating. We propose that planted trees may largely be chosen for specific traits (ecosystem service-based traits) rather than for species composition per se. Our third objective was to evaluate whether residents' preferences for specific tree attributes are reflected in the composition of the urban tree community, for example, whether there are more shade trees in hotter neighborhoods. An understanding of these patterns can contribute to general theories of species assemblages in cultivated gardens and planted urban forests.

9.2 METHODS

9.2.1 STUDY LOCATION AND DATA COLLECTION

Urban forest composition was inventoried across three southern California counties: Los Angeles, Orange, and Riverside in 2010 and 2011. Within each county 12 or 13 neighborhoods (37 in total) were selected to span a range of income and age of development within each county, which were determined using historical records and census tract data (Figure 1;

Supplementary Table 1). Neighborhood boundaries were determined from local on-line sources (Supplementary Table 1), and on average there were 7.5 census tracts per neighborhood (Supplementary Table 1). In Los Angeles, all neighborhoods where trees were surveyed were in the city of Los Angeles, while in Orange and Riverside counties, the neighborhoods where trees were surveyed spanned 11 and 4 cities, respectively. In Riverside, seven of the neighborhoods were in the City of Riverside. In each neighborhood, ten 22.4 m in diameter circular plots (0.04 hectare) were randomly placed. Data was collected according to the protocol in the Urban FOREst Effects (UFORE) model, now iTree Eco (USDA, 2011). This involves counting and identifying each tree and the overall tree cover of each plot. Land use type (i.e., park, residential, commercial) was also classified according to the iTree criteria and each tree was designated as either a street tree or not. Of the species recorded, seven were shrubs or vines that were trained to grow as trees and were excluded from our analyses (for example, *Bougainvillea glabra*).

9.2.2 SOCIO-ECONOMIC AND ENVIRONMENTAL VARIABLES

The plots were geocoded (ArcGIS 9.2, ESRI, 2006) and overlaid with environmental and socio-economic variables. Climate variables (temperature, annual maximum; precipitation, average annual) were acquired from the PRISM Climate Group at Oregon State University at a 1 km pixel size (Corvallis, OR, 2012) and averaged over a 30 year period (1981–2010). Population density at the tract level were taken from the 2010 U.S. census demographic profile 1 (DP1). Median family income, year the homes were built and proportion of the population with a college degree or higher were taken at the block group level from the American Community Survey from 2006–2011 (United States Census Bureau, 2012). The average of all plots in a neighborhood were used for all subsequent analyses. Across all counties, the neighborhoods spanned a range of both environmental and socio-economic factors (Table 1), none of which were correlated with one-another (Supplementary Figure 1), except income and education. Overall, Riverside was hotter and drier than Los Angeles and Orange counties (Table 1); both Orange and Riverside were more recently developed than Los

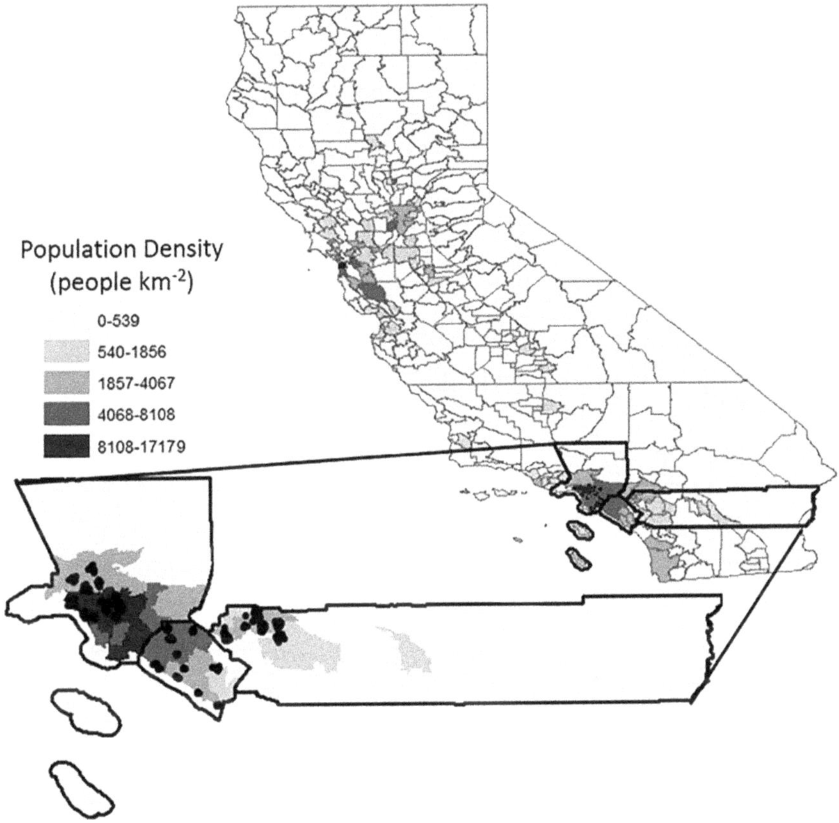

FIGURE 1: Locations of sampled plots across the 37 neighborhoods and three counties. The plots are clustered in neighborhoods.

Angeles; Orange County had a higher income than Riverside; and there was the higher population density in Los Angeles (Table 1).

9.2.3 ECOSYSTEM SERVICE-BASED TRAITS AND CLASSIFICATIONS

Based on a previous survey of residents in southern California, we focused on ecosystem service-based traits that were found to be important to residents (Pataki et al., 2013; Avolio et al., 2015; Supplementary Table 2).

Table 1. Differences among counties.

	F-value	P-value	LosAngeles (n = 13)	Orange (n = 12)	Riverside (n = 12)
# Trees $0.4ha^{-1}$	10.25	< 0.001	$19±1.7^{A}$	$24±2.6^{A}$	$12±1.3^{B}$
Tree Cover(%)	0.18	10.835	14±1.7	15±1.6	14±1.7
Tree Richness	6.19	0.005	$9±0.62^{AB}$	$12±1.15^{A}$	$8±0.71^{B}$
Precipitation (mm)	45.86	< 0.001	$417±13.7^{A}$	$336±9.4^{B}$	$269±8.8^{C}$
Temperature (C)	23.06	< 0.001	28±0.9B	$27±0.9^{B}$	$34±0.1^{A}$
Income (USD)	3.18	0.054	$87±965, 12, 783^{AB}$	$116±769, 14, 374^{A}$	$74±044, 7962^{B}$
Year homes built	5.70	0.007	$1961±2.7^{B}$	$1974±4.1^{A}$	$1978±4.7^{A}$
Population density (people km^{-2})	3.84	0.031	$5033±1153^{A}$	$3301±717^{AB}$	$1833±254^{B}$
Education	2.206	0.126	0.704±0.05	0.800±0.06	0.640±0.04

Shown are the F- and P-values from One-Way ANOVAs and the mean ± S.E. for each response variable. Letters indicated significant differences at $p \leq 0.05$ among the counties as determined by Tukey's HSD. Education is the proportion of the population with a college degree or higher.

Trait data for each tree species were collected from three sources: University of Florida's horticultural database (http://hort.ifas.ufl.edu/database/trees/trees_scientific.shtml), California Polytechnic State University's Urban Forest Ecosystems Institute database (http://selectree.calpoly.edu/), and from Sunset's Western Garden Book (Brenzel, 2001). Traits were classified at the species rather than the individual tree level. For example, a species that will provide a high degree of shade at maturity was counted as a shade tree, regardless of its current size and how much shade it actually provided. We utilized this method with the assumption that trees are likely procured and planted for their advertised traits, usually at maturity, and younger trees when they are planted might not yet possess the desired traits. See Supplementary Table 2 for a list of the ecosystem service-based traits used in this paper, traits are bolded and italicized here. The *provision of shade* had three categories with three being the highest shading potential. *Flowering* had three categories; 0 for species that did not flower (i.e., coniferous trees) or species with inconspicuous flowers (i.e., maple trees), (1) for trees whose flowers are visible (i.e., citrus trees) and (2) for trees that had large showy flowers (i.e., *Jacaranda* trees). We used two categories for *fruiting* depending on whether the species provides an edible product. We also used two categories for *fall color* depending on the presence of showy fall foliage. We used two categories of *fruit showiness* that depended on whether the fruit or berries are very visible. Overall, we derived a general category of "*showiness*" as an integrative measure of beauty. We calculated overall showiness as the sum of the flowering, fall color and showy fruit categories. Higher numbers indicated that a tree was more visually distinctive. *Growth rate* had three categories: $1 \leq 38$ cm $year^{-1}$, 2 = 39–76 cm $year^{-1}$, and $3 \geq 77$ cm $year^{-1}$. *Water requirement* was derived from the Western Garden Book (Brenzel, 2001) with a scale of 0–3. Zero was little to no water needs, (1) was used for species reported to require less than regular watering (every 2–3 weeks), (2) was used for species reported to need regular watering, and (3) was for species reported to require wet soils. Most of the species in this study were 1–3. *Damaging roots* were quantified with three categories depending on the degree to which roots could cause damage to the yard and sidewalk. The tendency to *drop litter* was quantified with two binary categories as well as NA for species for which there was insufficient information. The *phenology of leaves* was categorized as deciduous, evergreen or semi-evergreen. We also derived an index of *tree maintenance* as the sum of fruiting, high water requirement (water requirement > 2), dropping of debris, and deciduousness which ranged from 0 to 4. *Native* and *palm* both had two categories depending on whether the species was native to southern California or a palm species, respectively. Lastly, *leaf color*, *leaf type* (broad, needle, scaly), and *flower color* were determined, leaf color did not take into account if the leaf changed colors in the fall. Trees that did not flower or have visible flowers had NA for flower color.

9.2.4 TREE TYPES

We used the Calflora (www.calflora.org) database to determine tree species (as defined by Calflora) that can regenerate naturally and have been recorded in Los Angeles, Orange, and Riverside counties (136 native species and 113 exotic species). The Calflora does not consider elevation, and species that are native only to higher elevation areas are still considered native when sampled at lower elevations. Any species we recorded that was not in the Calflora database we considered cultivated and not able to regenerate naturally in southern California.

9.2.5 DATA ANALYSIS

All data analyses were conducted in R (R Core Development Team, Vienna, Austria) and statistical significance was considered at $\alpha = 0.05$. Environmental and socio-economic data were averaged across all plots at the neighborhood level. Tree data from each plot were summed at the neighborhood level because the iTree plots were small and many contained only 1 tree species. Thus, we considered all 10 iTree plots as necessary to adequately sample a single neighborhood, and the unit of replication was the neighborhood. The components of the urban forest were measured three ways: (1) all trees in all land use types, (2) street trees only across all land use types, and (3) non-street trees in plots where the land use was classified residential or multi-family. These three categories are hereafter referred to as "all tree," "street tree," and "residential tree" data, respectively. We calculated tree richness, Shannon's diversity, evenness (Shannon's diversity/ log (species richness) and Whittaker's Index for beta diversity in the Vegan package (Oksanen et al., 2013). Three neighborhoods had no street trees and two neighborhoods had no residential trees; thus these were excluded from the appropriate analyses. Tree cover was summed across all 10 plots and divided by 1000, and thus is percent cover.

For traits, we evaluated the proportion of trees in each neighborhood that had specific traits (e.g., proportion of trees that provided high shade). We quantified the proportion of trees that had the highest shading potential (3), the highest water requirement (3), the presence of visible and recognizable flowers (1 or 2), the fastest growth rate (3), the most destructive root systems (3), a showiness variable of > 1, and maintenance value > 2. All other traits were binary. We performed stepwise multiple regressions with residential trees only to assess the degree to which expressed preferences of residents matched attributes of the urban forest. We performed a functional trait analysis to determine the functional dispersion of each neighborhood with dbFD in the FD package (Laliberté and Legendre, 2010), which takes into account both the dissimilarity of traits as well as the abundance of each species. For these analyses, we used Gower dissimilarity to determine how much neighborhoods differed in tree traits. Gower allows for both continuous and categorical traits to be analyzed simultaneously, thus all traits were included in these analyses.

One-Way ANOVAs were used to determine differences among counties in their urban forest, environmental and socio-economic variables. Two-Way ANOVAs were used to determine whether there were differences between street and residential trees across the three counties for both measures of diversity and traits. We used Tukey's HSD for all post-hoc testing. To study relationships between environmental and socio-economic drivers with the urban forest we performed forward and backwards stepwise multiple regressions using the MASS package (Venables and Ripley, 2002). We used the relaimpo package to calculate partial regression coefficients (Groemping, 2006). Although income and education were correlated, they were never both selected for inclusion in a final multiple regression model, which eliminated any problems of collinearity.

We tested for multivariate differences in street and residential trees as well as street trees in the different counties based on species composition and trait data with non-metric multidimensional scaling using metaMDS in the Vegan package (Oksanen et al., 2013). For neighborhood differences based on species composition we used Bray-Curtis dissimilarity, while for traits we used Euclidian distance based on the proportion of trees with a particular trait. Using the adonis function in the Vegan package, we performed permutational multivariate ANOVA to test whether the patterns of community and trait dissimilarity were significant. Lastly, to determine the relationship between geographic distance among neighborhoods and tree community similarity based on both presence/absence species data and traits, we performed Mantel correlations using the Vegan package. Geographic distances were calculated in ArcGIS using the measure tool. Species composition differences among neighborhoods were calculated using Jaccard dissimilarity and trait differences (all proportional traits) were calculated using Euclidean distance. We performed Mantel correlations for all trees, street trees, and residential trees.

9.3 RESULTS

9.3.1 SOUTHERN CALIFORNIA'S URBAN FOREST

Overall we found 114 trees species in the surveyed neighborhoods. Using the classifications provided by Calflora, we found that of these species 7% were native, 46.5% were exotic but can regenerate naturally, and 46.5% were exotics that cannot reproduce naturally and must be planted and maintained by residents. We found 64 tree species in Los Angeles County, 75 in Orange County and 45 in Riverside County. All counties had similar beta diversity (~0.835) and demonstrated high turnover in species across neighborhoods within a county. The most common tree species was Mexican fan palm (*Washingtonia robusta*) followed by queen palm (*Arecastrum romanzoffianum*) and Mediterranean cypress (*Cupressus sempervirens*). Of these three most common species, *W. robusta* and *C. sempervirens* can reproduce naturally, while *A. romanzoffianum* cannot. However, most of the trees we encountered were planted by residents or the city and were not growing spontaneously. We found that the most common tree species varied by county as well as tree type (street or residential; Figure 2). Only one of the most common species was native, *Quercus agrifolia*. There were also differences among counties where Orange and Los Angeles counties had a greater number of trees per neighborhood than Riverside County, and neighborhoods in Orange County had greater tree richness than Riverside (Table 1). We also found some differences in land use type where residential areas had the greatest number of tree species and tree cover while utility areas had the fewest number of species and vacant lots had the least tree cover (Table 2).

9.3.2 DETERMINANTS OF URBAN FOREST STRUCTURE AND RICHNESS

Overall, socio-economic drivers had a greater effect on urban forest structure than environmental drivers, with neighborhood income, year the neighborhood was built, and proportion of residents with a college degree

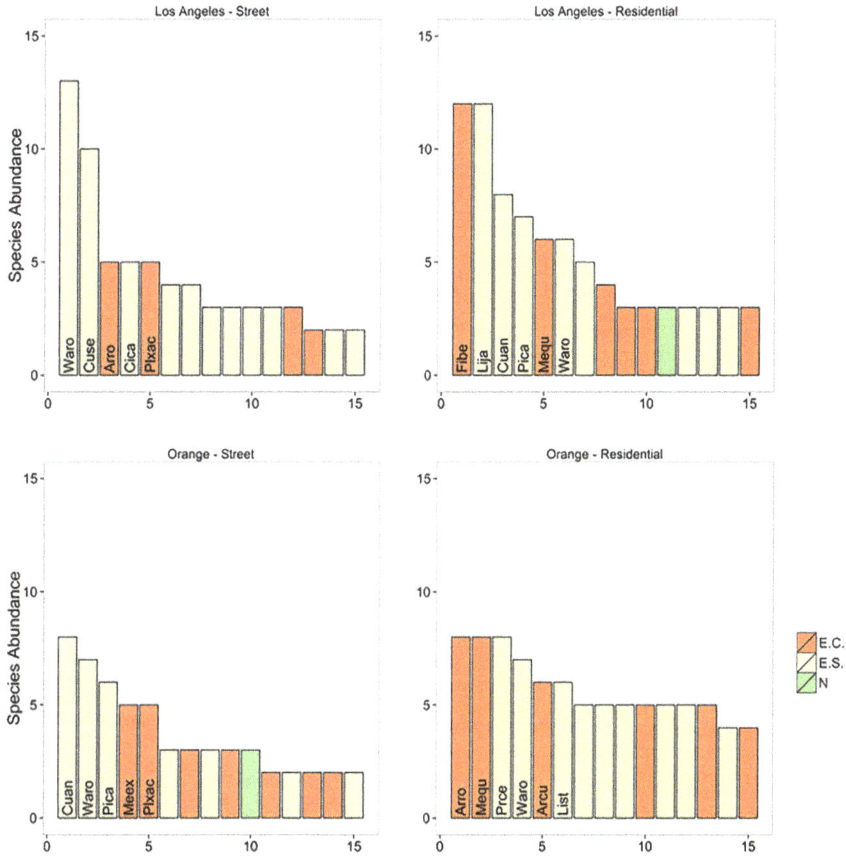

FIGURE 2: Rank abundance curves for the 15 most common street and residential trees across the three counties. Also shown are whether the species is native (N), is exotic but can reproduce naturally or spontaneously (ES), or is an exotic species that must be cultivated (EC). The names of most common 5–7 species are provided. Note that although many of the trees are exotic but can reproduce spontaneously, the majority of trees were still planted. This is especially true for the street trees. Species codes: *Archontophoenix cunninghamiana* (Arcu); *Arecastrum romanzoffianum* (Arro); *Cinnamomum camphora* (Cica); *Cupaniopsis anacardioides* (Cuan); *Cupressus sempervirens* (Cuse); Ficus benjamina (Fibe); *Koelreuteria bipinnata* (Kobi); *Ligustrum japonicum* (Lija); *Liquidambar styraciflua* (List); *Magnolia grandiflora* (Magr); Melaleuca quinquenervia (Mequ); Metrosideros excelsus (Meex); Pinus canariensis (Pica); *Platanus x acerifolia* (Plxac); *Podocarpus gracilior* (Pogr); *Prunus cerasifera* (Prce); *Pyrus calleryana* (Pyca); *Quercus agrifolia* (Quag); *Quercus ilex* (Quil); *Ulmus parvifolia* (Ulpa); *Schinus molle* (Scmo); *Washingtonia robusta* (Waro).

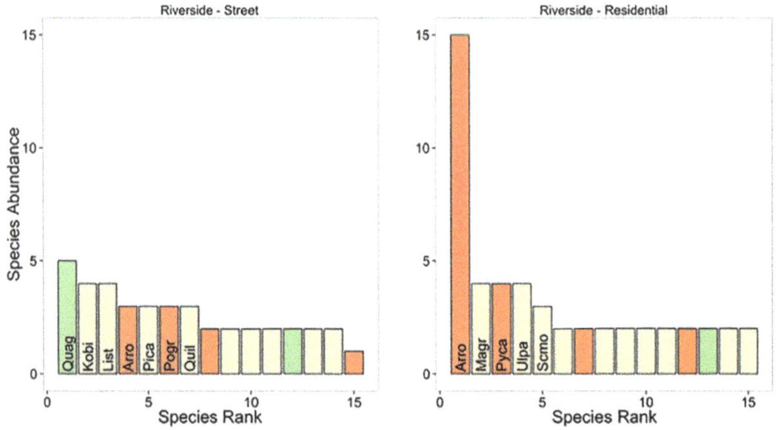

FIGURE 2: Continued.

or higher being the most important (Table 3). We also found that drivers differed depending on scale, either across all counties or within counties (Table 3). There were more trees in wealthier neighborhoods (Figure 3, Table 2) across all counties and Orange County, and there were also more trees in older neighborhoods in Orange County (Figure 3, Table 3). There was greater tree cover in more educated neighborhoods (Figure 3, Table 3) across all counties, and within Riverside County there was greater cover in older neighborhoods (Figure 3, Table 3). Across all counties and in Orange County alone there was greater richness in wealthier neighborhoods, and in Orange County only there was greater richness in older neighborhoods (Figure 3, Table 3). In Los Angeles County only, there was greater tree richness in neighborhoods where residents were more educated (Figure 3, Table 3). Across all counties, functional dispersion was not explained by any environmental or socio-economic drivers, but was negatively related to income in Riverside County only (Figure 3, Table 3).

9.3.3 PATTERNS OF URBAN FOREST DIVERSITY, COMPOSITION, AND TRAITS

When only looking at street and residential trees across the three counties, we found an interaction between county and tree type for tree richness

Table 2. Patterns of tree community diversity by land-use type across three counties in southern California.

County	Landuse	N	Alpha diversity			Beta diversity	Gamma diversity	# of stems	% Tree cover
			Richness	Evenness	H′				
All	Commercial	55	1.16 (0.07)	0.105 (0.04)	0.094 (0.04)	0.595 (0.01)	21	0.78 (0.18)	3.63 (0.79)
All	Institutional	19	1.16 (0.12)	0.105 (0.07)	0.094 (0.07)	0.464 (0.04)	7	0.79(0.32)	10.26 (4.09)
All	Park	14	1.64 (0.22)	0.393 (0.13)	0.358 (0.12)	0.942 (0.02)	16	3.34(1.16)	33.93 (7.54)
All	Transport.	13	1.23 (0.17)	0.148 (0.10)	0.133 (0.09)	0.628 (0.06)	7	2.69 (1.31)	10.38 (4.29)
All	Utility	5	1.4(0.24)	0.356 (0.22)	0.247 (0.15)	0.700 (0.15)	4	1.80 (1.11)	13.00 (9.69)
All	Vacant	25	1.08 (0.05)	0.073 (0.05)	0.051 (0.03)	0.300 (0.03)	6	0.32 (0.17)	2.80 (1.44)
All	Multi-family	42	1.67 (0.15)	0.390 (0.07)	0.351 (0.07)	0.930 (0.01)	37	2.53 (0.37)	15.83 (1.93)
All	Residential	188	1.54 (0.06)	0.366 (0.03)	0.302 (0.03)	0.960 (0.001)	90	2.18 (0.16)	18.37 (1.16)
LA	Residential	56	1.52 (0.11)	0.322 (0.06)	0.271 (0.05)	0.962 (0.004)	44	2.30 (0.38)	19.01 (2.03)
OC	Residential	66	1.73 (0.10)	0.510 (0.06)	0.418 (0.05)	0.969 (0.003)	60	2.65 (0.25)	16.28 (1.82)
RI	Residential	66	1.36(0.08)	0.260 (0.05)	0.211 (0.05)	0.948 (0.004)	39	1.62 (0.17)	19.92 (2.14)

N is the number of iTree plots that were classified as each land use type. Please note that this is not at the neighborhood scale but considers each iTree plot separately.Beta diversity was measured as the Whittaker Index.Shown are means and standard error in parentheses of all the plots in use land use class.

Table 3. Relationships of between characteristics of the urban forest for all trees with socio-economic and environmental drivers.

	Model Adj. R^2 (AIC)	Temp.	Precip.	Pop. Den.	Income	Yr. Built	Edu.
NUMBER OF TREES							
Los Angeles	n.s.						
Orange	0.732** (40.1)	0.044	0.028		0.345**	0.411**	
Riverside	n.s.						
All counties	0.302** (146.6)			0.082*	0.259**		
TREE COVER							
Los Angeles	n.s.						
Orange	n.s.						
Riverside	0.778** (−83.1)	0.151		0.144		0.523**	
All counties	0.236** (−211.2)	0.122				0.052	0.124*
RICHNESS							
Los Angeles	0.509* (15.1)	0.063	0.054				0.514**
Orange	0.762* (19.6)	0.087*	0.051	0.024	0.084*	0.624**	
Riverside	n.s.						
All counties	0.261** (82.8)			0.039	0.184**	0.100	
FUNCTIONAL DISPERSION							
Los Angeles	n.s.						
Orange	n.s.						
Riverside	0.506** (−93.8)	0.075	0.086		0.493*		
All counties	n.s.						

Models were run for each county separately and for all counties together. If an explanatory factor was not included in the final model the cell is left blank. Otherwise, relative importance of the factor is reported. For the significant explanatory variables, a cell shaded dark gray is a positive relationship and light gray is a negative relationship between the explanatory factor and measures of the urban tree community. For the overall model, and relative importance of individual factors significance is shown as: *$p < 0.05$, **$p < 0.01$.

(Table 4; Figure 4A), where in both Los Angeles and Riverside there was similar tree richness between street and residential trees, but in Orange County there was greater tree richness of residential trees compared with street trees. There was an effect of county for proportion of shade trees (Table 4), where there were more shade trees in Riverside compared with

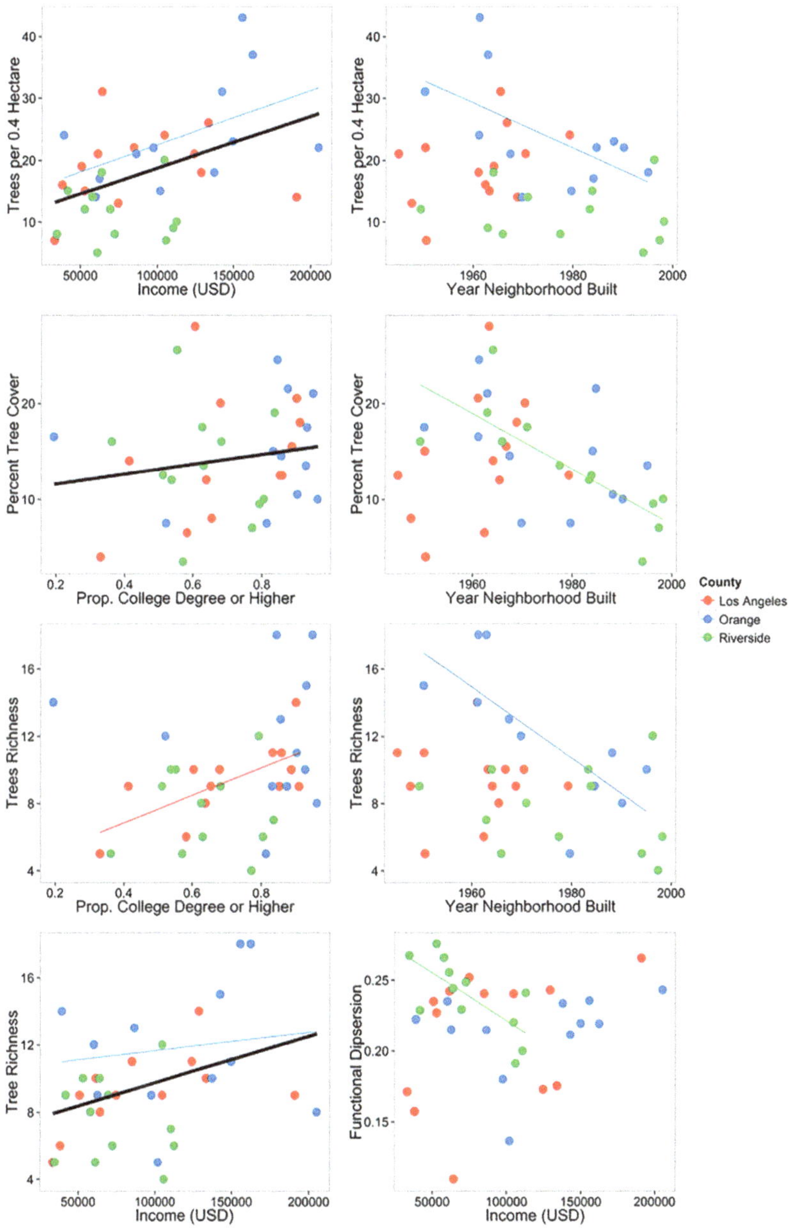

FIGURE 3: Overall patterns of urban forest characteristics in urbanized southern California. Data are for all neighborhoods across the three counties. A regression line is only shown for significant relationships. Black thick lines are the relationships across all counties, and thinner gray-scale lines are the relationships within each county.

Table 4. Differences between counties and tree type (street or residential) in community diversity and traits.

Factor	d.f.	Rich.	Even.	Div.(H')	Func. Disp.	Shade	Fruit	Flow.	Showy	Fast gr. rt.	Dam. rts.	High water req.	High maint.
County	2	3.45 (0.04)	2.36 (0.10)	1.88 (0.16)	1.78 (0.18)	4.92 (0.01)	2.54 (0.09)	1.25 (0.29)	0.10 (0.91)	1.43 (0.25)	2.92 (0.06)	2.60 (0.08)	1.77 (0.17)
Type	1	8.11 (0.01)	0.83 (0.37)	3.29 (0.07)	1.08 (0.30)	4.18 (0.05)	4.94 (0.03)	2.27 (0.14)	0.32 (0.57)	1.01 (0.31)	2.18 (0.15)	14.2 (<0.01)	5.63 (0.02)
C x T	2	3.48 (0.04)	0.74 (0.48)	5.14 (0.01)	0.91 (0.41)	2.19 (0.12)	1.87 (0.16)	0.72 (0.49)	1.50 (0.23)	0.15 (0.86)	1.92 (0.15)	0.38 (0.69)	1.35 (0.35)

Shown are the F- and P-values from Two-Way ANOVAs. Significant values are bolded.

Los Angeles (data not shown). We found more differences between tree types (street and residential trees) than among counties based on their traits (Table 4). A greater proportion of street trees provided shade compared with residential trees (Table 4; Figure 4B), while a greater proportion of residential trees provided fruit compared with street trees (Table 4). We also found that a greater proportion of residential trees had high water requirements (Table 4; Figure 4C) and a greater proportion had higher maintenance needs (Table 4; Figure 4D) compared with street trees.

There were no differences in species composition (Figure 5A; $p = 0.488$) between street and residential trees, although we were able to detect overall trait differences between street and residential trees (Figure 5C; $p = 0.009$). Similarly, we tested for differences among counties in street tree composition. For both species composition (Figure 5B; $p = 0.133$) and traits (Figure 5D; $p = 0.534$) there were no differences in street trees among counties.

Neighborhoods that were closer together did not have more similar tree communities for all trees ($r = 0.0336$, $p = 0.224$) and residential trees only ($r = 0.0144$, $p = 0.350$). For street trees only, we found that neighborhoods that were close together did have more similar tree species composition ($r = 0.0774$, $p = 0.050$). There were no distinguishable patterns in trait similarity and distance among neighborhoods for all trees ($r = -0.0345$, $p = 0.749$), street trees ($r = -0.0697$, $p = 0.994$), or residential trees ($r = -0.018$, $p = 0.593$).

9.3.4 LINKS BETWEEN RESIDENTIAL PREFERENCES AND RESIDENTIAL TREES COMMUNITIES

We found more residential trees in higher income neighborhoods (Table 5) and that temperature alone explained 26% of variation in the proportion of shade trees, where hotter neighborhoods had a greater proportion of shade trees (Table 5; Figure 6). There was a higher proportion of fruiting trees in older neighborhoods, more flowering trees in neighborhoods where residents were more educated, and a lower proportion of showy trees in neighborhoods with higher population densities (Table 5). For undesirable traits, we found a greater proportion of trees with damaging roots in hotter,

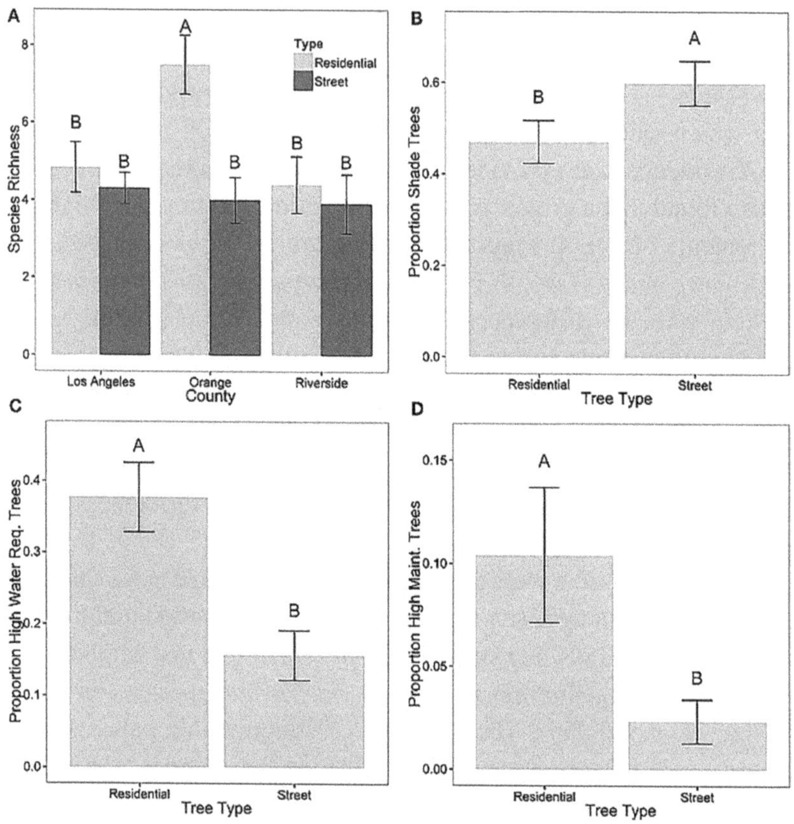

FIGURE 4: Patterns of community diversity across counties and tree types in species richness (A), proportion of shade trees (B), proportion of trees with high water requirements (C), and proportion of trees that are high maintenance (D). Shown are means ± standard error. Letters denote significant differences $p < 0.05$.

drier neighborhoods (Table 5). Lastly, we found a lower proportion of high maintenance trees in newer neighborhoods (Table 5).

9.4 DISCUSSION

Urban ecosystems are increasing in area worldwide (Grimm et al., 2008) and yet, we have relatively little ecological theory to understand what

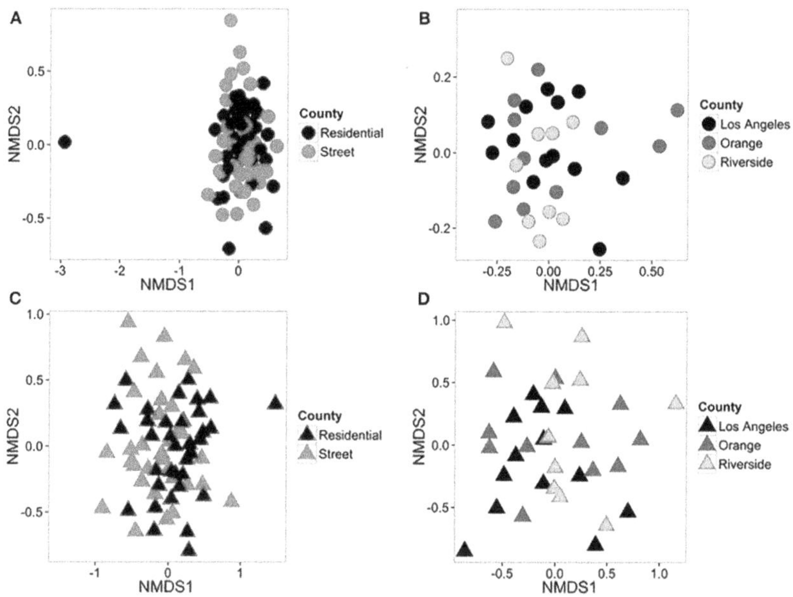

FIGURE 5: Differentiation of tree types across counties (street and residential) and counties (street trees only). We were unable to differentiate between street and residential trees (A, stress = 0.149), or between counties (B, stress = 0.104) using species composition data. Using trait data, we were able to detect differences between street and residential trees (C, stress = 0.198), but not counties using all traits (D, stress = 0.157). Species data is shown in circles and trait data in triangles.

controls population and community processes in urban forests. The current lack of understanding is associated with the complex and varied ways residents shape urban plant communities. In semi-arid and arid cities, urbanization increases the number of trees overall, and this pattern appears to be quite generalizable (McBride and Jacobs, 1976; Zipperer et al., 1997; Jenerette et al., 2013). Given that a large proportion of trees in these cities are planted, it seems reasonable to assume that sociological drivers are important determinants of richness and that attributes of these trees should be related to the preferences and management concerns of the actors who select trees. Here, we found that a new set of ecosystem service-based traits are very useful for understanding drivers of urban forest composition.

Table 5. Relationships of residential trees' functional dispersion, number of trees, and proportion of trees with a particular trait and socio-economic and environmental drivers.

	Model Adj. R^2 (AIC)	Temp.	Precip.	Pop. Den.	Income	Yr. Built	Edu.
Functional dispersion	n.s.						
Number of trees	0.159* (121.7)			0.035	0.173**		
Shade trees	0.264** (−97.9)	0.285**					
Fruiting trees	0.198** (−113.4)	0.078		0.019		0.183*	0.030
Flowering trees	0.091* (−93.7)						0.118*
Showy trees	0.170** (−93.9)	0.034		0.185**			
Trees with damaging roots	0.283** (−120.75)	0.159**	0.158**		0.034	0.016	
High water needs trees	n.s.						
Fast growing trees	n.s.						
High maintenance trees	0.109* (−116.7)					0.135*	

If an explanatory factor was not included in the final model the cell is left blank. Otherwise, the significance p-value of the factor is reported. For the significant explanatory variables, a cell shaded dark gray is a positive relationship and light gray is a negative relationship between the explanatory factor and measures of the urban tree community.For the overall model, and relative importance of individual factors significance is shown as: $*p < 0.05$, $**p < 0.01$.

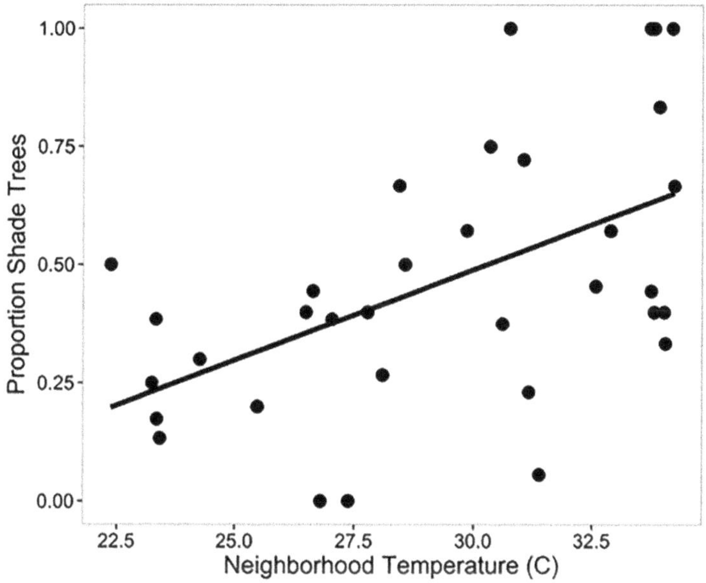

FIGURE 6: The relationship between the proportion of trees that provide significant shade and local neighborhood temperature of residential trees. See Table 4 for significance.

9.4.1 SOUTHERN CALIFORNIA'S URBAN FOREST

We found over 10 fold more exotic species than native species in southern California. Southern California is naturally an ecosystem where trees were not a dominant feature of the landscape at low elevations prior to urbanization (Rundel and Gustafson, 2005), and a general trend of more exotic tree species in urban areas has been found elsewhere (Aronson et al., 2015). The vegetation of southern California was surveyed in the 1980's (Miller and Winer, 1984) and street trees were surveyed in 1990's (Lesser, 1996). In 1984, the most common trees were California fan palm (*Washingtonia filifera*), Italian cypress (*Cupressus sempervirens*), and Monterey pine (*Pinus radiata*; Miller and Winer, 1984). All three species were found in our study; however, now the most common species is the exotic Mexican fan palm (*W. robusta*) and Monterey pine was not very common. In 1996, the most common street trees were American sweetgum (*Liquidambar styraciflua*), southern magnolia (*Magnolia grandifolia*), and holly oak (*Quercus ilex*; Lesser, 1996), and the most recently planted street trees were American sweetgum; crape myrtle (*Lagerstroemia indica*), and London planetree (*Platanus x acerifolia*; Lesser, 1996). Of the most commonly planted species in 1996 both American sweetgum and London planetree were among the five most common species. Many of the species that dominated our survey in 2010 were not as common 20–30 years ago, which may reflect the changing nature of resident preferences and species that are available in nurseries, which do change over time (Pincetl et al., 2013).

9.4.2 DETERMINANTS OF URBAN FOREST STRUCTURE AND RICHNESS

In contrast to Kendal et al. (2012b), who found that biophysical factors explained patterns of richness, we found that only socio-economic factors were significant, which is similar to patterns in Phoenix, AZ (Hope et al., 2003). One possible reason for this is the location of this research in southern California. Southern California has historically been described as a "Garden of Eden" (Pincetl et al., 2013). The temperatures are mild and with rampant irrigation urban plants are not reliant on rainwater. This is one possible reason why we did not find an effect of precipitation; trees are heavily irrigated in southern California (Pataki et al., 2011). The effect of climate and environmental conditions is most likely much stronger in cities where temperatures are more extreme.

Similar to other studies, we found greater species richness in higher income neighborhoods (Hope et al., 2003; Martin et al., 2004), although to our knowledge this is the first study to find this pattern when focusing on trees only. In a previous study that focused on the city of Los Angeles, Clarke et al. (2013) found that tree diversity was interactively affected by both development age and household income, where older, wealthier neighborhoods had the highest richness and new low income neighborhoods had the lowest richness. In southern California, maintenance costs of trees include irrigation (Pataki et al., 2011), which may enable wealthier areas to have greater tree cover (Jenerette et al., 2013). Income can also result in greater richness through complex social interactions, termed the "ecology of prestige" (Grove et al., 2006, 2014), whereby homeowner's desire to create an aesthetic that is associated with wealth. In addition to income, we found a marginally significant effect of neighborhood age ($p = 0.06$) with higher diversity in older neighborhoods. In Orange County, however, neighborhood age was the best predictor of richness and explained over 60% of variation in species richness. In Los Angeles County we found education was the best predictor or richness, explaining 62% of variation in richness. To our knowledge we are the first study to link education to biodiversity, other studies have found an effect of education on tree cover only (Heynen and Lindsey, 2003; Luck et al., 2009; Kendal et al., 2012b). In Los Angles, income was not included in the final multiple-regression model, suggesting that the education effect is not that these residents had more money. Instead, our findings suggests that education could be affecting ones attitude toward the importance of urban forests and trees in general, resulting in more pro-environmental behavior (Chen et al., 2011). Lastly, we did not find an effect of population density, perhaps because most neighborhoods were characterized by relatively high population density (Figure 1) and thus there was not much variation in this driver.

We also found the scale at which we were looking, within or across counties, affected drivers of the urban forest. For example, we were unable to detect a relationship between tree richness and socio-economic or environmental variables in Riverside County. This suggests that similar to natural areas, the scale at which diversity is assessed can be an important determinant of the associated drivers (Whittaker et al., 2001; Field et al., 2009). Hence, at the county scale there is reduced range of values to correlate with diversity relative to the regional scale, similar to findings in natural ecosystems (Field et al., 2009) in which drivers of diversity are more difficult to detect at smaller spatial scales. For example, there was a reduced range of neighborhood ages in Los Angles compared with Riverside and Orange Counties. Further, while we chose neighborhoods to span a range of income and age, we only sampled 10 neighborhoods per county, and perhaps a more exhaustive sampling would have revealed more patterns.

9.4.3 PATTERNS OF URBAN FOREST DIVERSITY, COMPOSITION, AND TRAITS

We found differences among counties, land use types, and street and residential trees in multiple measures of community diversity. Among land use types we found the highest species richness in parks and residential properties and the lowest in vacant lots and intuitional properties (Table 2). We also found differences between street and residential trees and an interaction with county, where there was greater richness of residential trees in Orange County but not in Los Angeles and Riverside Counties. Jim (1993) also found greater residential tree diversity than street tree diversity within a neighborhood of Hong Kong. Overall, we found different levels of species richness depending on land use types and tree types (residential vs. street), which may reflect the many different managers and drivers of the urban forest.

We postulate that residents may be less concerned with individual species, with which they are often unfamiliar, and more concerned with the functional and visual attributes that species provide. Trait identification and classification may be more informative than species identity and richness. When we compared traits of residential and street trees we found that residential areas had a greater proportion of fruiting trees and street trees had lower water requirements and needed less maintenance, but there was no significant difference in species composition between street and residential trees. Taken together, residents may plant trees more for provisioning ecosystem services while city managers are more concerned with tree water requirements and maintenance.

Neighborhoods that were closer together had more similar street tree species than neighborhoods farther apart. Although patterns of spatial auto-correlation are commonly found in natural ecosystems (Koerner and Collins, 2013), previous studies on urban vegetation did not find evidence of spatial auto-correlation (Hope et al., 2003; Clarke et al., 2013). We only found evidence of spatial auto-correlation with street trees, not all trees or residential trees, which could be caused by similar planting choices within municipalities or neighborhoods closer together may have been developed at similar times and reflect the planting preferences of that time period.

Overall, our findings suggest that neighborhoods within cities might have more similar tree communities than neighborhoods in different cities.

9.4.4 LINKS BETWEEN RESIDENTIAL PREFERENCES AND RESIDENTIAL TREES COMMUNITIES

Avolio et al. (2015) surveyed of preferences of residents across five southern California counties, including Los Angeles, Orange, and Riverside. This survey included residents in the neighborhoods studied here, however, we are not able to link preferences of residents in those specific neighborhoods with traits of trees in their neighborhood due to limitation of the dataset. Based on previous findings about the preferences of urban residents in southern California for specific tree attributes, we hypothesized that these preferences would shape the traits of the urban forest. For example, wealthier residents in southern California ranked the importance of having trees in their yard more highly than lower income residents (Avolio et al., 2015), and correspondingly we found more residential trees in higher income neighborhoods than lower income neighborhoods. Residents that lived in hotter areas had a greater preference for shade trees (Avolio et al., 2015), and here we found more shade trees in hotter neighborhoods. This is contrary to patterns expected based on biophysical drivers of forest processes alone, in that leaf area generally declines with increasing temperature (Cornelissen et al., 2003). Hence, planting preferences may completely overcome biophysical drivers and limitations. Although residents in more arid areas had a greater preference for trees that required less water (Avolio et al., 2015), we did not find that drier neighborhoods had more drought tolerant trees; tree watering requirements were not explained by any of the independent variables. This may be partially explained by the low cost of water in southern California, such that irrigation requirements may not have historically played an important role in decision-making about trees. Similarly, although provision of fruit was more important to low income than high-income residents (Avolio et al., 2015), here we found that the proportion of fruit trees was not related to income. A possible cause of this mismatch between resident preferences and traits of the urban forest is that monetary limitations may restrict the ability of lower income residents to create tree communities that match their preferences. By comparing stated preferences with patterns of urban forest diversity and traits we found that resident preferences are reflected in traits of the urban forest, and that the strength of this relationship may be modified by resident's income.

9.5 CONCLUSION

In southern California we found a diverse urban forest primarily composed of exotic species. Overall, socio-economic variables better explained variation in species richness, number of trees, tree cover, and functional dispersion than environmental variables. Additionally, we found that within county drivers of the urban forest were not the same at larger geographic scales, highlighting the need for ecologists to study the scale at which drivers of urban diversity are most influential. We found linkages between resident preferences for specific tree attributes and the actual distribution of functional traits in the urban forest. For example, we found that residents in hotter neighborhoods have a greater tendency to prefer shade trees, and currently this is manifest in more shade trees in hotter neighborhoods. Overall our results show that the majority of tree species in the urbanized region of southern California are exotic species, about half of which need to be actively planted by humans for their survival. As such, the attributes or traits that residents and managers use to select which species to plant are key to understanding patterns of urban vegetation.

REFERENCES

1. Anderson, K. (2007). Temporal patterns in rates of community change during succession. Am. Nat. 169, 780–793. doi: 10.1086/516653
2. Aronson, M. F. J., Handel, S. N., La Puma, I. P., and Clemants, S. E. (2015). Urbanization promotes non-native woody species and diverse plant assemblages in the New York metropolitan region. Urban Ecosyst. 18, 31–45. doi: 10.1007/s11252-014-0382-z
3. Avolio, M. L., Pataki, D. E., Pincetl, S., Gillespie, T. W., Jenerette, G. D., and McCarthy, H. R. (2015). Understanding preferences for tree attributes: the relative

effects of socio-economic and local environmental factors. Urban Ecosyst. 18, 73–86. doi: 10.1007/s11252-014-0388-6
4. Boone, C. G., Cadenasso, M. L., Grove, J. M., Schwarz, K., and Buckley, G. L. (2009). Landscape, vegetation characteristics, and group identity in an urban and suburban watershed: why the 60s matter. Urban Ecosyst. 13, 255–271. doi: 10.1007/s11252-009-0118-7
5. Bourne, K. S., and Conway, T. M. (2013). The influence of land use type and municipal context on urban tree species diversity. Urban Ecosyst. 17, 329–348. doi: 10.1007/s11252-013-0317-0
6. Brenzel, K. N. (ed). (2001). Sunset Western Garden Book, 7th Edn. Menlo Park, CA: Sunset Published Co.
7. Chen, X., Peterson, M. N., Hull, V., Lu, C., Lee, G. D., Hong, D., et al. (2011). Effects of attitudinal and sociodemographic factors on pro-environmental behaviour in urban China. Eniron. Conserv. 38, 45–52. doi: 10.1017/S037689291000086X
8. Clarke, L. W., Jenerette, G. D., and Davila, A. (2013). The luxury of vegetation and the legacy of tree biodiversity in Los Angeles, CA. Landsc. Urban Plan. 116, 48–59. doi: 10.1016/j.landurbplan.2013.04.006
9. Cook, E. M., Hall, S. J., and Larson, K. L. (2012). Residential landscapes as social-ecological systems: a synthesis of multi-scalar interactions between people and their home environment. Urban Ecosyst. 15, 19–52. doi: 10.1007/s11252-011-0197-0
10. Cornelissen, J. H. C., Lavorel, S., Garnier, E., Díaz, S., Buchmann, N., Gurvich, D. E., et al. (2003). A handbook of protocols for standardised and easy measurement of plant functional traits worldwide. Aust. J. Bot. 51, 335–380. doi: 10.1071/BT02124
11. Díaz, S., Cabido, M., and Casanoves, F. (1998). Plant functional traits and environmental filters at a regional scale. J. Veg. Sci. 9, 113–122. doi: 10.2307/3237229
12. Field, R., Hawkins, B. A., Cornell, H. V., Currie, D. J., Diniz-Filho, A. F., Guégan, J.-F., et al. (2009). Spatial species-richness gradients across scales: a meta-analysis. J. Biogeogr. 36, 132–147. doi: 10.1111/j.1365-2699.2008.01963.x
13. Gaston, K. J. (2000). Global patterns in biodiversity. Nature 40, 220–227. doi: 10.1038/35012228
14. Grimm, N. B., Faeth, S. H., Golubiewski, N. E., Redman, C. L., Wu, J., Bai, X., et al. (2008). Global change and the ecology of cities. Science 319, 756–760. doi: 10.1126/science.1150195
15. Groemping, U. (2006). Relative importance for linear regression in r: the package relaimpo. J. Stat. Softw. 17, 1–27.
16. Grove, J. M., Locke, D. H., and O'Neil-Dunne, J. P. M. (2014). An ecology of prestige in New York City: examining the relationships among population density, socio-economic status, group identity, and residential canopy cover. Environ. Manage. 54, 402–419. doi: 10.1007/s00267-014-0310-2
17. Grove, J. M., Troy, A. R., O'Neil-Dunne, J. P. M., Burch, W. R. Jr., Cadenasso, M. L., and Pickett, S. T. A. (2006). Characterization of households and its implications for the vegetation of urban ecosystems. Ecosystems 9, 578–597. doi: 10.1007/s10021-006-0116-z

18. Heynen, N., and Lindsey, G. (2003). Correlates of urban forest canopy cover: implications for local public works. Public Works Manag. Policy. 8, 33–47. doi: 10.1177/1087724X03008001004
19. Hope, D., Gries, C., Zhu, W., Fagan, W. F., Redman, C. L., Grimm, N. B., et al. (2003). Socioeconomics drive urban plant diversity. Proc. Natl. Acad. Sci. U.S.A. 100, 8788–8792. doi: 10.1073/pnas.1537557100
20. Iverson, L., and Cook, E. (2000). Urban forest cover of the Chicago region and its relation to household density and income. Urban Ecosyst. 4, 105–124. doi: 10.1023/A:1011307327314
21. Jenerette, G. D., Harlan, S. L., Brazel, A., Jones, N., Larissa, L., and Stefanov, W. L. (2007). Regional relationships between surface temperature, vegetation, and human settlement in a rapidly urbanizing ecosystem. Landsc. Ecol. 22, 353–365. doi: 10.1007/s10980-006-9032-z
22. Jenerette, G. D., Miller, G., Buyantuev, A., Pataki, D. E., Gillespie, T. W., and Pincetl, S. (2013). Urban vegetation and income segregation in drylands: a synthesis of seven metropolitan regions in the southwestern United States. Environ. Res. Lett. 8:044001. doi: 10.1088/1748-9326/8/4/044001
23. Jim, C. Y., and Chen, W. Y. (2008). Pattern and divergence of tree communities in Taipei's main urban green spaces. Landsc. Urban Plan. 84, 312–323. doi: 10.1016/j.landurbplan.2007.09.001
24. Jim, C. Y., and Liu, H. T. (2001a). Patterns and dynamics of urban forests in relation to land use and development history in Guangzhou City, China. Geogr. J. 167, 358–375. doi: 10.1111/1475-4959.00031
25. Jim, C. Y., and Liu, H. T. (2001b). Species diversity of three major urban forest types in Guangzhou City, China. For. Ecol. Manage. 146, 99–114. doi: 10.1016/S0378-1127(00)00449-7
26. Jim, C. Y. (1993). Trees and landscape of a suburban residential neighbourhood in Hong Kong. Landsc. Urban Plan. 23, 119–143. doi: 10.1016/0169-2046(93)90112-Q
27. Kendal, D., Williams, K. J. H., and Williams, N. S. G. (2012a). Plant traits link people's plant preferences to the composition of their gardens. Landsc. Urban. Plan. 105, 34–42. doi: 10.1016/j.landurbplan.2011.11.023
28. Kendal, D., Williams, N. S. G., and Williams, K. J. H. (2012b). Drivers of diversity and tree cover in gardens, parks and streetscapes in an Australian city. Urban For. Urban Green. 11, 257–265. doi: 10.1016/j.ufug.2012.03.005
29. Kirkpatrick, J. B., Daniels, G. D., and Davison, A. (2011). Temporal and spatial variation in garden and street trees in six eastern Australian cities. Landsc. Urban Plan. 101, 244–252. doi: 10.1016/j.landurbplan.2011.02.029
30. Koerner, S. E., and Collins, S. L. (2013). Small-scale patch structure in North American and South African grasslands responds differently to fire and grazing. Landsc. Ecol. 28, 1293–1306. doi: 10.1007/s10980-013-9866-0
31. Kunick, W. (1987). Woody vegetation in settlements. Landsc. Urban Plan. 14, 57–78. doi: 10.1016/0169-2046(87)90006-5
32. Laliberté, E., and Legendre, P. (2010). A distance-based framework for measuring functional diversity from multiple traits. Ecology 91, 299–305. doi: 10.1890/08-2244.1

33. Lesser, L. (1996). Street tree diversity and DBH in Southern California. J. Arboric. 22, 180–186.
34. Lowry, J. H., Baker, M. E., and Ramsey, R. D. (2011). Determinants of urban tree canopy in residential neighborhoods: Household characteristics, urban form, and the geophysical landscape. Urban Ecosyst. 15, 247–266. doi: 10.1007/s11252-011-0185-4
35. Luck, G. W., Smallbone, L. T., and O'Brien, R. (2009). Socio-economics and vegetation change in urban ecosystems: patterns in space and time. Ecosyst. 12, 604–620. doi: 10.1007/s10021-009-9244-6
36. Maco, S., and McPherson, E. (2002). Assessing canopy cover over streets and sidewalks in street tree populations. J. Arboric. 28, 270–277.
37. Martin, C. A., Warren, P. S., and Kinzig, A. P. (2004). Neighborhood socioeconomic status is a useful predictor of perennial landscape vegetation in residential neighborhoods and embedded small parks of Phoenix, AZ. Landsc. Urban Plan. 69, 355–368. doi: 10.1016/j.landurbplan.2003.10.034
38. McBride, J., and Jacobs, D. (1976). Urban forest development: a case study, Menlo Park, California. Urban Ecol. 2, 1–14. doi: 10.1016/0304-4009(76)90002-4
39. McGill, B. J., Enquist, B. J., Weiher, E., and Westoby, M. (2006). Rebuilding community ecology from functional traits. Trends Ecol. Evol. 21, 178–185. doi: 10.1016/j.tree.2006.02.002
40. Miller, P., and Winer, A. (1984). Composition and dominance in Los Angeles Basin urban vegetation. Urban Ecol. 8, 29–54. doi: 10.1016/0304-4009(84)90005-6
41. Oksanen, J., Blanchet, F. G., Kindt, R., Legendre, P., Minchin, P. R., O'Hara, R. B., et al. (2013). vegan: Community Ecology Package. R package version 2.0-9. Available online at: http://CRAN.R-project.org/package=vegan
42. Pataki, D., McCarthy, H., Gillespie, T., Jenerette, G. D., and Pincetl, S. (2013). A trait-based ecology of the Los Angeles urban forest. Ecosphere 4:art72. doi: 10.1890/ES13-00017.1
43. Pataki, D. E., McCarthy, H. R., Litvak, E., and Pincetl, S. (2011). Transpiration of urban forests in the Los Angeles metropolitan area. Ecol. Appl. 21, 661–677. doi: 10.1890/09-1717.1
44. Pearce, L. (2013). Using size class distributions of species to deduce the compositional dynamics of the private urban forest. Arboric. Urban For. 39, 74–84.
45. Pincetl, S., Gillespie, T., Pataki, D. E., Saatchi, S., and Saphores, J.-D. (2012). Urban tree planting programs, function or fashion? Los Angeles and urban tree planting campaigns. GeoJournal 78, 475–493. doi: 10.1007/s10708-012-9446-x
46. Pincetl, S., Prabhu, S. S., Gillespie, T. W., Jenerette, G. D., and Pataki, D. E. (2013). The evolution of tree nursery offerings in Los Angeles County over the last 110 years. Landsc. Urban Plan. 118, 10–17. doi: 10.1016/j.landurbplan.2013.05.002
47. Reich, P. B., Walters, M. B., and Ellsworth, D. S. (1997). From tropics to tundra: global convergence in plant functioning. Proc. Natl. Acad. Sci. U.S.A. 94, 13730–13734. doi: 10.1073/pnas.94.25.13730
48. Roman, L. A., and Scatena, F. N. (2011). Street tree survival rates: meta-analysis of previous studies and application to a field survey in Philadelphia, PA, USA. Urban For. Urban Green. 10, 269–274. doi: 10.1016/j.ufug.2011.05.008

49. Rundel, P. W., and Gustafson, R. (2005). Introduction to the Plant Life of Southern California. University of California Press, Los Angeles.
50. Sanders, R. (1984). Some determinants of urban forest structure. Urban Ecol. 8, 13–27. doi: 10.1016/0304-4009(84)90004-4
51. Sjöman, H., Östberg, J., and Bühler, O. (2012). Diversity and distribution of the urban tree population in ten major Nordic cities. Urban For. Urban Green. 11, 31–39. doi: 10.1016/j.ufug.2011.09.004
52. Talarchek, G. (1990). The urban forest of New Orleans: an exploratory analysis of relationships. Urban Geogr. 11, 65–86. doi: 10.2747/0272-3638.11.1.65
53. Vallet, J., Daniel, H., Beaujouan, V., Rozé, F., and Pavoine, S. (2010). Using biological traits to assess how urbanization filters plant species of small woodlands. Appl. Veg. Sci. 13, 412–424. doi: 10.1111/j.1654-109X.2010.01087.x
54. Venables, W. N., and Ripley, B. D. (2002). Modern Applied Statistics with S. New York, NY: Springer.
55. Whittaker, R. J., Willis, K. J., and Field, R. (2001). Scale and species richness: towards a general, hierarchical theory of species diversity. J. Biogeogr. 28, 453–470. doi: 10.1046/j.1365-2699.2001.00563.x
56. Williams, N. S., Schwartz, M. W., Vesk, P. A., McCarthy, M. A., Hahs, A. K., Clemants, S. E., et al. (2009). A conceptual framework for predicting the effects of urban environments on floras. J. Ecol. 97, 4–9. doi: 10.1111/j.1365-2745.2008.01460.x
57. Zhang, H., and Jim, C. Y. (2014). Species diversity and performance assessment of trees in domestic gardens. Landsc. Urban Plan. 128, 23–34. doi: 10.1016/j.landurbplan.2014.04.017
58. Zipperer, W., Sisinni, S., Pouyat, R., and Foresman, T. (1997). Urban tree cover: an ecological perspective. Urban Ecosyst. 1, 229–246. doi: 10.1023/A:1018587830636

CHAPTER 10

Assessing the Effects of the Urban Forest Restoration Effort of MillionTreesNYC on the Structure and Functioning of New York City Ecosystems

P. TIMON MCPHEARSON, MICHAEL FELLER, ALEXANDER FELSON, RICHARD KARTY, JACQUELINE W.T. LU, MATTHEW I. PALMER, AND TIM WENSKUS

10.1 INTRODUCTION

Urban areas are complex combinations of ecological remnants with varying states of human development. Urbanized areas cover only 1% to 6% of Earth's surface, yet they have massive ecological footprints (Rees and Wackernagel 1996) and complex and often indirect effects on surrounding ecosystems (Alberti et al. 2003). Urbanized land already covers more area than the combined total of national and state parks and areas preserved by The Nature Conservancy (McKinney 2002). Urban areas continue to expand as populations increase. For example, New York City (NYC) expects to add nearly 1 million residents by 2030 to an already densely populated city. Additionally, 70% of all humans globally are predicted to live in cities by 2050 (US Census Bureau 2000). Given these trends, one of the primary dynamics that must be understood at a local, regional, and global scale is the effect of humans on the ecology of urban systems (Machlis et al. 1997; Pickett and Grove 2009).

The contemporary ecological paradigm recognizes that humans are integral parts of ecosystems exerting direct and indirect influences on the functioning of ecological systems (Egerton 1993; McDonnell and Pickett 1993; Holling 1994; Cronon 1995; Alberti et al. 2003; Turner et al. 2004). However, the study of urban ecosystems is still a relatively new pursuit in ecology (Pickett et al. 2001; Pickett and Grove 2009). The need to understand the intricacies of urban ecosystems emerges from the increasing fraction of humanity that calls cities home and from the disproportionate impact cities have on both regional and global systems (Collins et al. 2000; Grimm et al. 2000; Pickett et al. 2008). A more nuanced understanding of urban ecosystems, including socio-ecological dynamics, would allow ecologists to use socio-ecological theory to explain and predict urban dynamics (Pickett et al. 2008). Similarly, understanding of urban ecological patterns and processes would allow for improved, adaptive management of cities for healthier and more resilient socio-ecosystems.

There are a number of important examples of ecosystem research in NYC. Early groundwork for an understanding of cities as socio-ecological systems was laid by William H. Whyte's social ecology program in NYC (Whyte 1980; 1988) and continues to be developed by many others (Platt 2006). In addition, the urban to rural gradient studies developed two decades ago (McDonnell and Pickett 1990; McDonnell et al. 1997) and revisited over the years (Gregg et al. 2003) made significant contributions to urban ecology. Here we discuss an initial step towards a greater understanding of NYC as an urban ecosystem through a multi-institutional, interdisciplinary, long-term research study of the dynamics of urban forested ecosystems through the installation of long-term urban forest research plots across NYC.

10.2 PLANYC 2030 / MILLIONTREESNYC

On Earth Day 2007 NYC Mayor Michael Bloomberg announced PlaNYC, a comprehensive longterm sustainability plan for New York City (City of New York 2007). PlaNYC includes 127 ambitious sustainability initiatives, one of which is the MillionTreesNYC (MTNYC) Initiative, a public-private partnership between the NYC Department of Parks & Recreation (NYC Parks) and the New York Restoration Project (NYRP), with the goal of planting one million trees by 2017. Since the launch of MTNYC, public, private and non-profit organizations have organized nearly 4,000 citizen volunteers to plant trees across NYC and inspired planting campaigns in other U.S. cities. One aspect of MTNYC directs the planting of nearly 400,000 trees to establish 2,000 acres of new forest on NYC parkland and other public open spaces with the goal of creating multi-story, ecologically functioning forests. This large-scale afforestation effort provides the basis for a citywide ecological research project discussed here.

10.3 THE ECOLOGICAL VALUE OF URBAN FORESTS

Urban forests provide cities with numerous ecological benefits including: regulating local surface and air temperatures, filtering pollution from the local atmosphere which may positively impact the health of urban residents, trapping rainwater during heavy storms which prevents pollution of local waterways, and storing and sequestering atmospheric carbon dioxide. One recent study by the U.S. Forest Service put the compensatory value of NYC's forest at over $5 billion (Nowak at el. 2007) using the Urban Forest Effects Model (UFORE) and data collected in 1997 on the city's forest. UFORE estimated that NYC's forest stores 1.35 million tons of carbon, a service valued at $24.9 million. The forest sequesters an additional 42,300 tons of carbon per year (valued at $779,000 per year) and about 2,202 tons of air pollution per year (valued at $10.6 million per year; Nowak et al. 2007).

We suggest that increased information on the structure and functioning of the urban forest can be used to improve and augment support for urban forest management programs and to integrate urban forests within plans to improve environmental quality in the NYC area. Now in its third year, the city has already added over 300,000 young trees to existing urban parks, private lands, and city streets (Figure 1). But will planting trees result in the kinds of complex multi-story structures and ecological functioning desired of forests? How will various planting strategies affect these outcomes?

10.4 ESTABLISHING LONG-TERM EXPERIMENTAL PLOTS IN NYC

NYC Parks' Natural Resources Group (NRG) has a long history of coupling ecological research and monitoring with applied urban vegetation management and ecological restoration practices. This has included grant funding and collaboration with universities. For the MTNYC effort, NYC Parks in 2008 worked with EDAW | AECOM, a consulting firm, and with the MTNYC Advisory Board's Research and Evaluation Subcommittee to establish a large-scale research project designed as functional parkland (Felson and Pickett 2005). The goal was to study the short and long-term impacts of the MTNYC tree planting strategies on ecosystem structure and functioning in a couple key NYC parks. More recently, researchers joined with NYC Parks to develop a more comprehensive citywide research project. The project represents a partnership between NRG, The New School's Tishman Environment and Design Center (TEDC), Columbia University's Department of Ecology, Evolution, and Environmental Biology (E3B), and the Yale School of Forestry and Environmental Studies.

This research leverages the large-scale tree planting activities of the MTNYC campaign to create structured experimental study plot treatments in order to understand the effects of MTNYC's forest restoration efforts on the structure and functioning of urban parkland in NYC (Figure 2). We define forest restoration here as the cumulative management activities of invasive plant removal, dense tree and shrub planting, and soil amendment as motivated and designed by NYC Parks in parks citywide. Motivating questions for our research include: How do variations in planting practices affect the development trajectories of new forest communities? How long will it take for forest canopy closure under different management practices, and how does closure rate affect invasive plant population dynamics? How do planting decisions and restoration practices affect overall forest restoration success, as measured by canopy closure and rate of invasive plant establishment? What are the implications of expected heterogeneity in soil nutrients for plant dynamics and productivity and

FIGURE 1: MillionTreesNYC Tree Planting Since 2007. MillionTreesNYC plants trees in parks, privately held land, along streets, and other areas around the city with the goal of adding one million trees to NYC by 2017. Source: MillionTreesNYC.org

how might soils be in turn affected as the plant community develops? The goal of the research is to work towards understanding several of these key management questions through a multi-year study to provide baseline scientific data to inform park design and forest management. We will monitor survivorship and growth of individual trees and measure canopy density

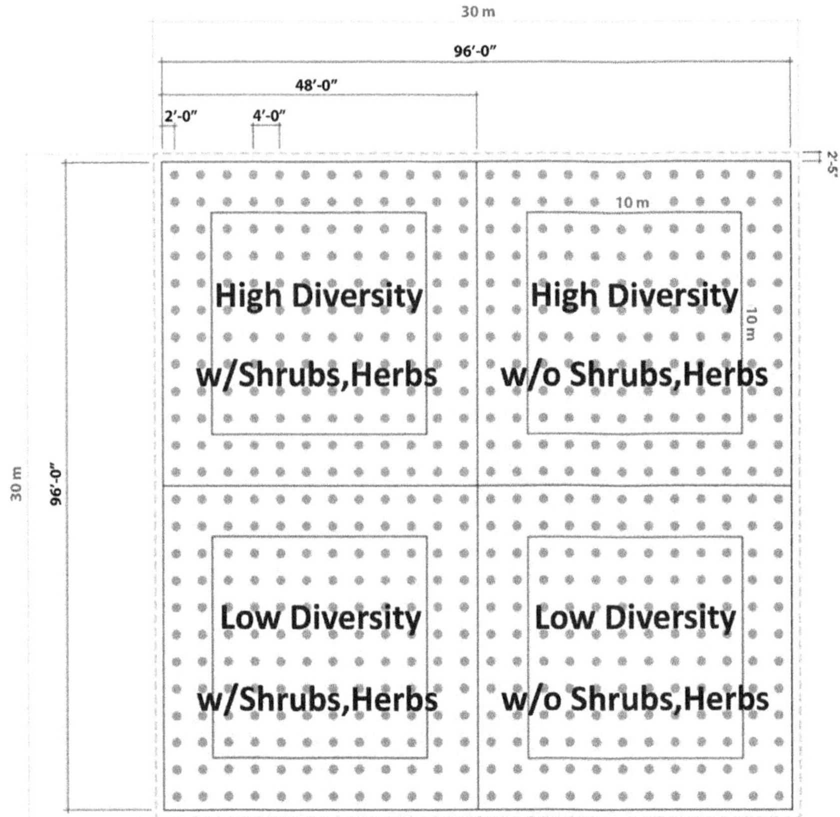

FIGURE 2: Permanent Plot Design. Experimental research plots consist of a 30m x 30m plot with four 15m x 15m nested subplots in a block design with two main treatments, High and Low Diversity and Understory (w/Shrubs, Herbs) or No Understory (w/o Shrubs, Herbs). Trees are planted four feet on center (shown as green dots). Vegetation and soil data is annually sampled in a 10m x 10m plot nested within the 15m x 15m subplots to minimize edge effects between subplot treatments. Diversity and understory treatments are randomly applied to the subplots when the plot is established. The arrangement of the subplots varies in some parks based on the size and shape of the area being restored.

at the stand level, as well as assess the understory vegetation and changes to soils, both as they exist at the initiation of the restoration and as they develop over time.

10.5 URBAN VEGETATION AND SOIL ANALYSIS

Long-term study of forest restoration and regeneration is critical to understanding forest dynamics in urban ecosystems. Urban vegetation and soil studies are important to understand urban biodiversity, climate modification, carbon dynamics, and pollution and water absorption functions of soil. We are particularly interested in the role of exotic and invasive species, which have received particular attention in urban ecology (Pickett et al. 2001). In an earlier urban-to-rural gradient study in NYC, the number of exotics in the seedling and sapling size classes of woody species was greater in urban and suburban oakdominated stands (Rudnicky and McDonnell 1989). There is growing evidence that the presence of exotics is enhanced along pathways in rural recreation areas (Rapoport 1993) and in urban parks (Drayton and Primack 1996). In Boston's Middlesex Fells, a 400 ha urban woodland park inventoried for plants in 1894, a re-census of the flora in 1993 showed that the majority of new species recorded on the site were exotic species and that native species had declined by nearly 10% (Drayton and Primack 1996). By studying vegetation in a large number of heterogeneous sites across the city, we hope to build a more comprehensive picture of invasive plant population dynamics and their effects on the ecological dynamics of NYC forests.

Understanding the ecological and management controls on plant diversity is critical for understanding how ecosystems function. The relationship between biodiversity and ecosystem functioning has been an area of intense debate in the ecological sciences (Naeem 2002). It has been argued from theory and empirically demonstrated that biodiversity should increase the functioning of ecosystems (Loreau et al. 2002; Cardinale et al. 2006). However, depending on the functional characteristic measured, this prediction has not held up in all empirical investigations of the relationship (Jiang et al. 2008). In urban ecosystems the question is even murkier. We are examining this relationship in a subset of afforested parks in NYC forest ecosystems by looking specifically at changes in diversity over time and the relationship between diversity, forest development, and ecosystem functions such as net primary productivity and soil carbon storage.

Assessing baseline and changing soil conditions is also essential for prioritizing further ecosystemscale research in urban forests and for understanding the impacts of soils on vegetation dynamics and restoration outcomes in urban areas. Urban soils are known to be highly heterogeneous (Pouyat et al. 2007). However, soils in NYC are poorly understood and a simultaneous investigation of both citywide (New York City Soil Survey 2005) and local, plot-scale soils will provide critical data for building a more comprehensive understanding of urban ecosystem dynamics. This research is designed to assess how soil heterogeneity varies across space and time in NYC's forested ecosystems and the effects of this heterogeneity on vegetation dynamics. This project will focus first on characterizing the heterogeneity within and among research plots, thereby providing data on variation in soil nutrients, metals, and carbon at local and regional scales. We are interested in whether soil heterogeneity within study plots impacts the survivorship and growth of trees, shrubs, and herbaceous species planted in the MTNYC campaign and whether heterogeneity across sites can help to explain potential variation in species performance.

Most soil studies in urban areas have focused on disturbed and human-constructed soils along streets and in highly developed areas (Craul and Klein 1980; Patterson et al. 1980; Short et al. 1986; Jim 1993, 1998; Pouyat et al. 2007). As a result "urban soils" typically have been viewed as drastically disturbed soil material of low fertility (Craul 1999). Yet other potentially influential factors associated with urban land transformations have received limited attention. In fact, the characteristics of soil can vary greatly across the urban landscape, including not only highly disturbed, but also relatively undisturbed soils that are modified by management and urban environmental factors (Schleuß et al. 1998; Pouyat et al. 2003). Urban soil research that describes the differences in surface soil properties among various land uses and cover types will be useful in differentiating relatively intact remnant soils from highly disturbed and managed soils, and for assessing impacts of soil on vegetation dynamics in long- term research plots. In addition, those soil properties associated with specific management strategies (such as those employed in MTNYC) and intensity of use may be useful as diagnostic properties to differentiate human impacts on surface soil characteristics in urban landscapes (Pouyat et al. 2007).

10.6 HYPOTHESES

We examine the dynamic interactions between plants, soils, and management practices in permanent forest restoration plots, focusing on how they change over time. This research is guided by three overarching hypotheses:

1) Forest restoration will enhance urban forest functioning (e.g., net primary productivity and soil fertility) over time at the plot scale and citywide.
2) Forest restoration will increase the biological diversity of urban forests over time at the plot scale and citywide.
3) Forest restoration will decrease the abundance and distribution of invasive species over time at the plot scale and citywide.

10.7 METHODS

Evaluating park planting and management designs requires experimental treatments that can be implemented across sometimes very different park settings with adequate replication. Study plots need to be large enough to capture relevant dynamics but small enough to fit into interstitial restoration areas in existing parks. Methodological approaches also require simplicity given the multiple participants, including researchers, volunteers, local community members, and NYC Parks personnel. Plot size also needs to be reasonably small to allow efficient sampling on an annual basis as the number of plots increases with time (plots and thus replication increase over time as more reforestation sites are designated by NYC Parks & Recreation). The plot size should also reflect the need for permanent plots to facilitate additional field studies and subsequent research projects while meeting the goals of the current study.

Research collaborators chose 900 m^2 plots (Figure 2), which are similar in scale to other forest studies such as the U.S. Forest Service Forest Inventory and Analysis Program (U.S. Forest Service 2007). Long-term experimental research plots utilize a nested design to allow scientists to evaluate the importance of varying levels of tree diversity and understory on reforestation dynamics. Research plots are a randomized complete block design with four 15m x 15m subplots nested within each 30m x 30m full plot with two treatments (High Tree Diversity/Low Tree Diversity and Understory/No Understory) in a factorial experimental design (Figure 2). Treatments are designed to test how varying levels of tree diversity combined with understory or no understory treatments affect long-term restoration outcomes. Within each subplot is a 10m x 10m sampling plot. The subplot is centered within the treatment plot to minimize edge effects. Therefore, all vegetation and soil sampling takes place at the 10m x 10m subplot scale. Subplot corners are marked with permanently installed rebar with GPS coordinates recorded at plot corners. Site selection for permanent plots was based on availability of forest restoration sites of appropriate size (large enough to accommodate a 900 m^2 research plot) and canopy openness (in order to limit variation caused by shading from mature trees).

Subplots are planted, in coordination with NRG field crew leaders, MTNYC personnel, volunteers, and contractors, with 7.6 L (2-gallon) container trees (tree height varies from 0.5 - 1.0meters) in high (6 species) and low (2 species) diversity treatment configurations randomized across blocks within the full plot. The diversity levels were chosen to span the range of tree species richness typically found in areas of similar size in existing NYC urban forests. The understory treatment contains 3.8 L (1-gallon) shrubs planted at a density of 36 shrubs per subplot (Figure 2). Tree and shrub species were chosen based on known or expected adaptations to particular urban park conditions, local biophysical characteristics of site type, availability from local nurseries, and park landscape design parameters in collaboration with NYC Parks ecologists to establish standardized planting palettes for both mesic and hydric site types across the city. Mesic sites include six tree species (*Quercus rubra*, *Nyssa sylvatica*, *Amelanchier canadensis*, *Prunus serotina*, *Quercus coccinea*, and *Celtis occidentalis*) and six shrub species (*Sambucus canadensis*, *Lindera benzoin*, *Aronia arbutifolia*, *Rosa virginiana*, *Viburnum acerifolium*, and *Hamamelis virginiana*). Hydric sites also include six tree species (*Quercus palustris*, *Nyssa sylvatica*, *Quercus bicolor*, *Liquidambar styraciflua*, *Platanus occidentalis*, and *Diospyros virginiana*) and six shrub species (*Cornus amomum*, *Clethra alnifolia*, *Viburnum dentatum*, *Rosa palustris*,

Cephalanthus occidentalis, and *Ilex verticillata*). Tree and shrub density follows the current planting practices of NYC Parks, where trees are planted with approximately 1.2 m (four foot) spacing and shrubs (Figure 2). The expectation is that as the canopy closes, invasive plants will be shaded out in a natural process of competition with native trees for light, nutrients and water.

An important part of this study involves recording recent management history on all study sites, which typically involves invasive plant removal (by chemical sprays, selective cutting, and mowing) as a site preparation strategy. Invasive removal is a critical but costly preparation for tree seedling establishment in urban parks often dominated by invasive plant species. Invasive species management has become an area of intense focus and expense for NYC Parks. Current urban invasives removed as part of forest restoration efforts include *Artemisia vulgaris* (mugwort), *Phragmites australis* (common reed), *Rosa multiflora* (multiflora rose), *Ampelopsis brevipedunculata* (porcelain berry), *Ailanthus altissima* (tree of heaven), *Fallopia japonica* (Japanese knotweed), and *Celastrus orbiculatus* (Asiatic bittersweet). Management practices may also include soil amendments such as mulching newly planted trees and watering during the susceptible periods of early tree establishment. This project includes extensive interaction with NYC Parks staff to document recent (past three years) and current management at the research sites in order to understand the ecosystem management practices which may affect the experimental response variables.

Annual monitoring of vegetation and soils in permanent field plots will allow us to accumulate a time series of vegetation and soil dynamics data in order to follow community development among experimental treatments. Pilot research plots were installed in April 2009 to refine the experimental design and data collection methodology discussed above. The pilot sites were also vital to developing research protocols to coordinate NYC Parks' site preparation and management practices and MTNYC tree planting events with plot installation and data collection. Permanent long-term research plots were installed beginning in Summer 2009. Plot installation includes collecting pre-planting baseline vegetation and soil data. As of October 2010, permanent experimental plots have been established in the following parks: Roy Wilkins and Alley Pond in Queens; Clove

Lakes and Conference House in Staten Island; Pelham Bay in the Bronx; Canarsie and Marine in Brooklyn. We plan to add additional plots in subsequent years, expanding until MTNYC sites that meet the requirements of the research design are exhausted.

Plot scale analyses rely on both pre- and post-planting vegetation and soil assessment in order to monitor responses to experimental treatments. Annual vegetation and soil monitoring are completed in July and August (in order to maximize the potential to identify the largest proportion of plants within a single field visit), prior to scheduled forest restoration plantings in the fall (usually October). Vegetation data collection includes surveying trees, shrubs, and herbaceous plants at the 10m x 10m subplot scale at all sites. We sampled the presence and percent cover of all existing vegetation at the plot scale, which allows us to address questions about tree and shrub growth, regeneration and productivity, mortality, recruitment, density, invasive species dynamics, and other related metrics of vegetation structure and function. By examining tree, shrub, and herb dynamics over time, this project will establish the baseline database for further interdisciplinary analyses of other ecological, social, and economic impacts of forest restoration on urban ecosystems.

Tree and shrub cover is monitored using two line transects, 1cm wide by ~14.1m long, drawn diagonally from subplot corners, along which the total number of centimeters intercepted by individuals is recorded (Figure 3). The line intercept method has been used in other restoration studies in NYC parklands and has been used successfully in previous pilot studies with a high level of accuracy. The line intercept method is also used to assess the herbaceous plant community by stretching four 1cm wide x 10m long transects (H1-H4) one meter in from each subplot corner (Figure 3) and recording the total area that herbs intercept the line for a total of 4000 cm^2 cover per subplot. Shrubs are also assessed for cover and location, and size (dbh) of trees, if any, are recorded. Nearby canopy cover is measured using a spherical densiometer since trees near plots may impact light availability and therefore vegetation responses near plot edges. All vegetation measures are assessed annually and preliminary baseline vegetation results are presented below.

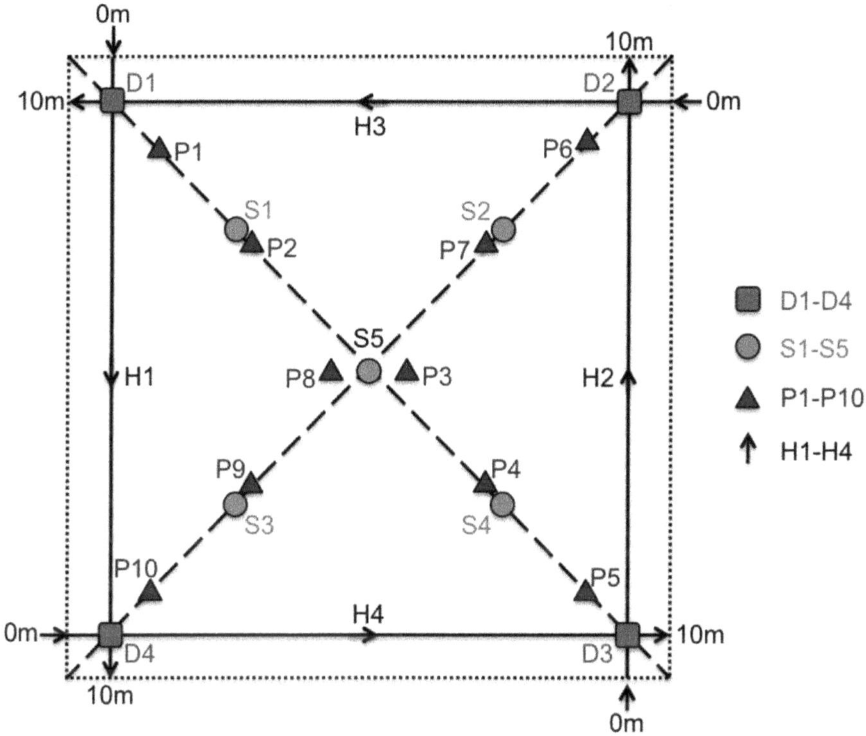

FIGURE 3: Subplot Annual Sampling Design. Vegetation and soil sampling occurs in each 10m x 10m subplots. D1-D4 refer to spherical densiometer measurements taken to assess canopy cover at each plot corner. P1-P10 are locations for high resolution soil samples, leaf litter measurements, and soil penetrometer readings taken every 2.36 meters along diagonals and twice offset from center. S1-S5 are locations for soil sampling locations used for composite samples. H1-H4 are transects used for percent cover assessment of vegetation including shrubs and herbs. *Tx* and *Ty* refer to diagonal transects used to establish soil sampling locations for each subplot.

We acquired soil samples within each subplot using two techniques: (i) by taking ten undisturbed 5-cm-diameter by 10-cm-deep samples (P1-P10; Figure 3) from one randomly chosen subplot in each full plot per site for high resolution soil analysis; and (ii) by taking a composite soil sample from 0-10cm depth, composited from 5 locations within each subplot (S1-S5; Figure 3), taken with a sampling probe. Soil monitoring in both pre- and post-planting phases includes assessing physical and chemical characteristics including all major nutrient, heavy metal, and carbon analyses, leaf litter depth, soil compaction, and pH. Subsamples of the pre-planted dry-sieved soil were analyzed for all major nutrients (P, K, Ca, Mg, Mn, Zn, Al, NO_3), total and organic carbon, and heavy metals (Al, Ca, Co, Cu, Cr, Fe, K, Mg, Mn, Na, Ni, P, Pb, S, Ti, V) at Cornell University's Nutrient Analysis Laboratory. However, initial soil results are not presented here. Fall 2009 site descriptions including general soil type descriptions are included in Appendix 1.

10.8 ANALYSES

Plant diversity and percent cover from plant data collected in July and August 2009 was analyzed from five of the six sites that were planted with MTNYC trees in October 2009 in Bronx, Queens, and Brooklyn (Alley Pond, Pelham Bay, Roy Wilkins, Marine and Canarsie Park). Clove Lakes Park in Staten Island was added late and was not sampled in 2009. For each site, species abundance (cm^2) was summed across all transects. Total abundance across all species for a single 1cm x 10m transect could exceed 1000 cm^2 because multiple species could occupy the same space as measured by vertical projection of the transect boundary (1cm x 10m) onto the ground. Proportions of introduced and invasive species at each site were calculated based on species counts, using nativity and invasive status information from USDA (2010) and Uva et al. (1997). Species coverage at each site was calculated by dividing each species' abundance by 16000 cm^2, the total sampled area at each site. Shannon diversity index (Shannon 1948) and evenness (Magurran 1988) were calculated using these cover values.

10.9 RESULTS AND DISCUSSION

Permanent research plots in seven different parks across Brooklyn, Queens, Bronx, and Staten Island have been installed and sampled to date. We show here preliminary vegetation diversity and percent cover results from the five permanent plots assessed in 2009. Plant diversity varied across the five sites with Shannon diversity highest in the Orchard Beach site in Pelham Bay Park, Bronx and species richness highest in Roy Wilkins Park, Queens (Table 1; see Appendix 2 for a complete species list). We have not yet investigated drivers of variation in species richness across our sites, though we expect site history and anthropogenic disturbance to be important. Similar studies along an urban-rural gradient in Germany found that non-native species richness was correlated with various indicators of anthropogenic disturbance, though native species richness was not (Brunzel et al. 2009).

Invasive species are of particular concern to forest restoration in NYC because of their ability to outcompete tree seedlings and, therefore, inhibit canopy development in MTNYC forest restoration sites. In initial analyses, all sites were dominated by invasive species prior to tree planting, with the highest proportion of vegetative cover by invasives in Marine Park, Brooklyn (91%), and the lowest in Canarsie Park, Brooklyn (71%; Figure 4). Interestingly, though all sites were dominated by invasive species, not all invasives were non-native. Initial surveys revealed that three sites (Alley Pond, Marine, Roy Wilkins Parks) dominated by non-native species, one site, Canarsie Park, had relatively equal cover of natives and non-natives, and the Pelham Bay site was dominated by natives, though the majority were still largely invasive (Figure 4).

The most abundant individual species in all study plots were invasive. For example, *Artemisia vulgaris* (Mugwort), a common non-native invasive (Barney et al. 2008, 2009) being combated in parks and private land throughout NYC, was the most abundant species in Canarsie, Marine, and Roy Wilkins Park, and second most abundant at Alley Pond Park. *Fallopia japonica* (Japanese knotweed) was the most abundant species in the Pelham Bay Park site. Rank abundance plots show the relatively steep curves for Alley Pond and Canarsie Park, which indicate how a small number of species dominate the sites, with abundance quickly dropping off among the lower-ranked, less-abundant species (Figure 5). Conversely, the less-steep curve in the rank abundance plot for Pelham Bay indicates a higher degree of evenness, with smaller differences between the more- and less-abundant species. We expect this research will provide direct measures of

Table 1. Diversity Across Fall 2009 Sites. Diversity is shown for Fall 2009 study sites (Alley Pond (1 plot), Canarsie (1 plot), Marine Park (calculated from 2 subplots), Pelham Bay (1 plot), Roy Wilkins (2 subplots)). Richness is total number of species found at a site. Diversity is Shannon Diversity, and evenness is calculated by dividing Shannon diversity by maximum possible diversity.

Site	Richness	Diversity	Evenness
Alley Pond	22	0.65	0.48
Canarsie Beach	23	0.69	0.51
Marine Park	19	0.81	0.63
Pelham Bay	28	0.95	0.66
Roy Wilkins	31	0.80	0.54

invasive species dynamics by linking plot scale data to a growing citywide analysis of the effect of management strategy on plant diversity and abundance. Additionally, aggregated vegetation data will allow an analysis of how community dynamics in different patches in the city change over time due to understory and tree diversity treatment variables.

10.10 CONCLUSION

The MTNYC reforestation experimental plots are a long-term project designed to understand the controls on urban ecosystem structure and function in forest restoration sites, and how ecosystem management practices may affect these controls. We focus on the abiotic and biotic drivers that may impact structure and function in urban forest vegetation and soils. The study is organized around repeated measurements of 900 m2 plots to provide a framework for scaling up in space and time. Study plots are located in parks throughout NYC (Appendix 3) and are sampled both before trees and understory species are planted, and annually thereafter in order to assess ecosystem change over time. Long-term study of urban ecosystems is critical to the future of urban ecology. Over the next several

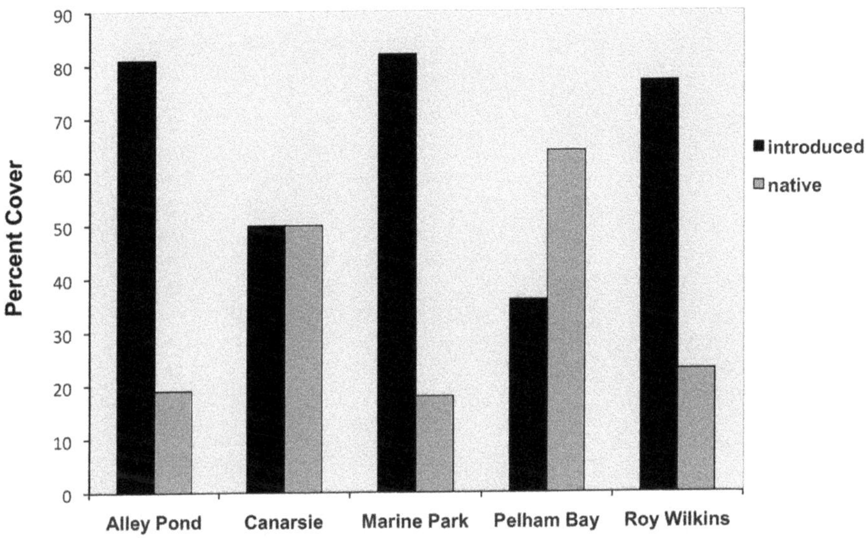

FIGURE 4A: The proportion of native and introduced plant abundance expressed as a percent of total abundance (y-axis)

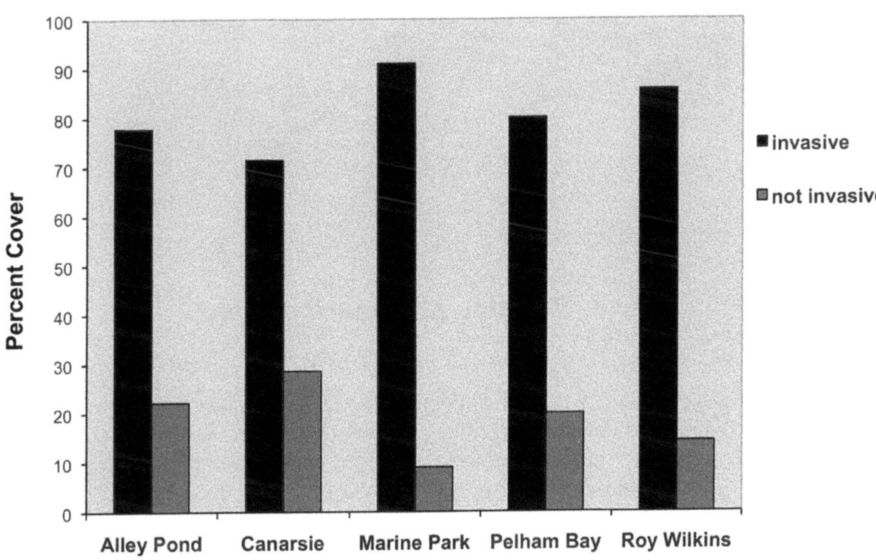

FIGURE 4B: The proportion of invasive and non-invasive plant abundance expressed as a percent of total abundance (y-axis). Abundance is calculated by measuring percent cover along transects in research plots in Alley Pond and Roy Wilkins Parks (Queens), Canarsie and Marine Parks (Brooklyn), and Pelham Bay Park (Bronx) (x-axis).

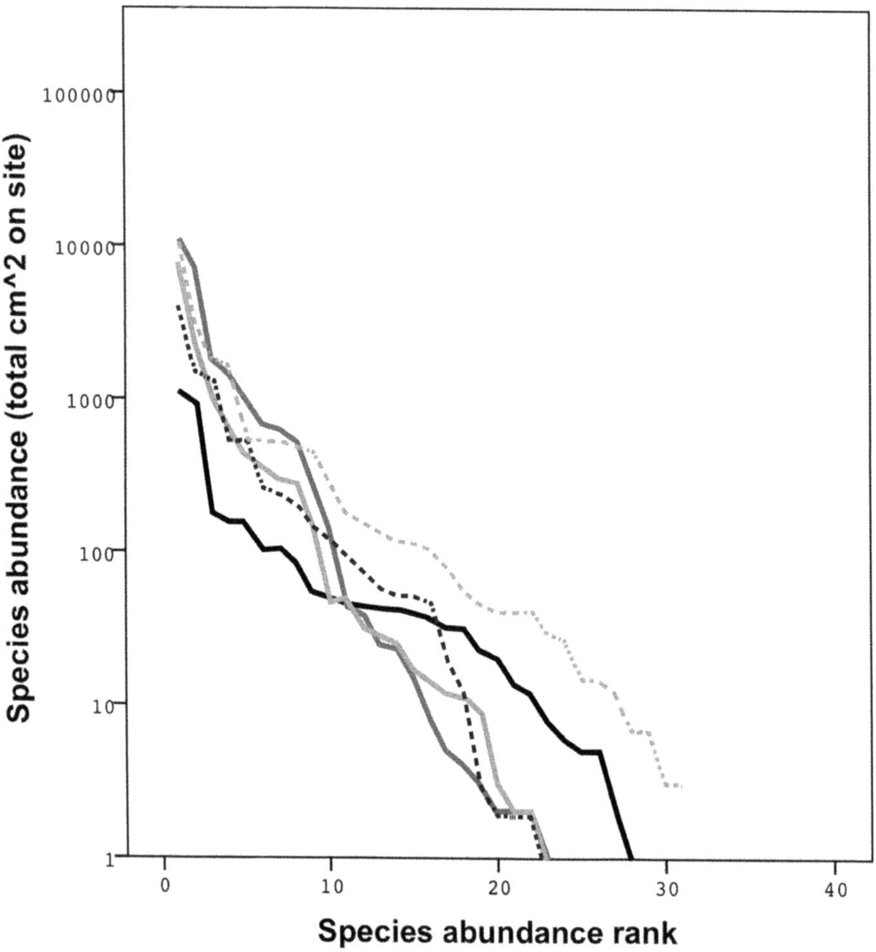

FIGURE 5: Rank Abundance for Fall 2009 Sites. The rank abundance plot shown with rank on the X-axis and abundance on the Y-axis for each site. Rank is a sequential number assigned to each species in decreasing order of abundance, within each site. For each site, the most-abundant species has a rank of one.

years, this project will focus on analyzing vegetation and soil data from the experimental research plots to better understand the development of urban forest ecosystems. This study will also provide a baseline of intensive data for future ecological research within NYC. We have found interesting vegetation patterns among sites across the city and expect with further analysis and integration of soil analyses to begin explaining these patterns. These analyses will provide new data for understanding the effects of 2000 acres of afforestation on ecosystem structure and functioning, and will provide the potential to connect intensive, neighborhood and site scale analyses of ecological, physical and social processes and mechanisms with other citywide, extensive research. Ultimately we expect our research on ecological restoration in urban centers and the impacts on the structure and functioning of regional scale environments to prove useful for urban ecosystem management and policymaking.

REFERENCES

1. Alberti, M., J.M. Marzluff, E. Shulenberger, G. Bradley, C. Ryan, and C. Zumbrunnen. 2003. Integrating humans into ecology: Opportunities and challenges for studying urban ecosystems. Urban Ecology 53:1169–1179.
2. Barney, J.N., T.H. Whitlow, A.J. Lembo. 2008. Revealing historic invasion patterns and potential invasion sites for two non-native plant species. PLoS ONE 3:1-8.
3. Barney, J.N., T.H. Whitlow, and A. DiTommaso. 2009. Evolution of an invasive phenotype: shift to belowground dominance and enhanced competitive ability in the introduced range. Plant Ecology 202:275–284.
4. Brunzel, S., S.F. Fischer, J. Schneider, J. Jetzkowitz and R. Brandl. 2009. Neo- and archaeophytes respond more strongly than natives to socio-economic mobility and disturbance patterns along an urban-rural gradient. Journal of Biogeography 36:825-844.
5. Cardinale B.J., D.S. Srivastava, J.E. Duffy, J.P. Wright, A.L. Downing, M. Sankaran, and C. Jouseau. 2006. Effects of biodiversity on the functioning of trophic groups and ecosystems. Nature 443:989-992.
6. Collins J.P., A. Kinzig, N.B. Grimm, W.F. Fagan, D. Hope, J. Wu, and E.T. Borer. 2000. A new urban ecology. American Scientist 88:416–425.
7. Craul, P.J. and C.J. Klein. 1980. Characterization of streetside soils of Syracuse, New York. p. 88–101. In METRIA 3: Proceedings of the Conference of the Metropolitan Tree Improvement Alliance, 3rd, Rutgers, NJ. 18–20 June 1980. North Carolina State University, Raleigh.
8. Craul, P.J. 1999. Urban soils: Applications and practices. John Wiley & Sons, New York.
9. Cronon, W. 1995. The trouble with wilderness; or, getting back to the wrong nature, pp. 69-90. In Cronon, W. (Ed.). Uncommon Ground: Toward Reinventing Nature. Norton, New York, NY.

10. Drayton B. and R.B. Primack. 1996. Plant species lost in an isolated conservation area in metropolitan Boston from 1894 to 1993. Conservation Biology 10:30–39.
11. Egerton, F.N. 1993. The history and present entanglements of some general ecological perspectives, pp 9-23. In McDonnell, M.J. and S.T.A. Pickett (Eds.). Humans as Components of Ecosystems: The Ecology of Subtle Human Effects and Populated Areas. Springer-Verlag, New York, NY.
12. Felson, A. and S.T.A. Pickett. 2005. Designed experiments: New approaches to studying urban ecosystems. Frontiers in Ecology & the Environment 3:549-556.
13. Gregg J.W.,C.G. Jones, and T.E. Dawson. 2003. Urbanization effects on tree growth in the vicinity of New York City. Nature 424:183-7.
14. Grimm, N.B., M. Grove, S.T.A. Pickett, and C.L. Redman. 2000. Integrated approaches to long-term studies of urban ecological systems. Bioscience 50: 571–584.
15. Holling, C.S. 1994. New science and new investments for a sustainable biosphere, pp. 57-97. In A. Jansson (Ed.). Investing in Natural capital: the Ecological Economics Approach to Sustainability. Island Press, Washington, D.C.
16. Jiang, L., P. Zhichao, and D. R. Nemergut. 2008. On the importance of the negative selection effect for the relationship between biodiversity and ecosystem functioning. Oikos 117: 488-493.
17. Jim, C.Y. 1993. Soil compaction as a constraint to tree growth in tropical and subtropical urban habitats. Environmental Conservation 20:35–49.
18. ------. 1998. Physical and chemical properties of a Hong Kong roadside soil in relation to urban tree growth. Urban Ecosystems 2:171–181.
19. Kühn, I., R. Brandl and S. Klotz. 2004. The flora of German cities is naturally species rich. Evolutionary Ecology Research 6:759-764.
20. Loreau, M., S. Naeem, and P. Inchausti (Eds.). 2002. Biodiversity and Ecosystem Functioning: Synthesis and Perspectives. Oxford University Press, Oxford. 312 pp.
21. Machlis, G.E., J.E. Force, and W.R. Burch, Jr. 1997. The human ecosystem. 1. The human ecosystem as an organizing concept in ecosystem management. Society and Natural Resources 10:347–367.
22. Magurran, A.E. 1988. Ecological diversity and its measurement. Princeton University Press, Princeton, NJ.
23. McDonnell, M. J. and S.T.A. Pickett. 1990. The study of ecosystem structure and function along urbanrural gradients: an unexploited opportunity for ecology. Ecology 71:1231–1237.
24. ------. 1993. Humans as components of ecosystems: the ecology of subtle human effects and populated areas. Springer-Verlag, New York, NY.
25. McDonnell, M. J., S.T.A. Pickett, R.V. Pouyat, R.W. Parmelee, M.M. Carreiro, P.M. Groffman, P. Bohlen, W.C. Zipperer, and K. Medley. 1997. Ecology of an urban-to-rural gradient. Urban Ecosystems 1:21-36.
26. McKinney, M.L. 2002. Urbanization, biodiversity, and conservation. BioScience 52:883–890.
27. Naeem, S. 2002. Ecosystem consequences of biodiversity loss: the evolution of a paradigm. Ecology 83:1537–1552.
28. New York City Soil Survey Staff. 2005. New York City Reconnaissance Soil Survey. United States Department of Agriculture, Natural Resources Conservation Service, Staten Island, NY.

29. Nowak, D. J., R.E. Hoehn, D.E. Crane, J.C. Stevens, and J.T. Walton. 2007. Assessing urban forest effects and values, New York City's urban forest. Resource Bulletin NRS-9. U.S. Department of Agriculture, Forest Service, Northern Research Station, Newtown Square, PA.
30. Patterson, J.C., J.J. Murray, and J.R. Short. 1980. The impact of urban soils on vegetation. p. 33–56. In METRIA 3: Proceedings of the Conference of the Metropolitan Tree Improvement Alliance, 3rd, Rutgers, NJ. 18–20 June 1980. North Carolina State University, Raleigh.
31. Pickett, S.T.A., M.L. Cadenasso, J.M. Grove, C.H. Nilon, R.V. Pouyat, W.C. Zipperer, and R.Costanza. 2001. Urban ecological systems: Linking terrestrial ecological, physical, and socioeconomic components of metropolitan areas. Annual Review of Ecology and Systematics 32:127–157.
32. Pickett, S.T.A., M.L. Cadenasso, J.M. Grove, P.M. Groffman, L.E. Band, C.G. Boone, W.R. Burch, C.S. B. Grimmond, J. Hom, J.C. Jenkins, N.L. Law, C.H. Nilon, R.V. Pouyat, K. Szlavecz, P.S. Warren, and M.A. Wilson. 2008. Beyond urban legends: An emerging framework of urban ecology, as illustrated by the Baltimore Ecosystem Study. Bioscience 58:139-150.
33. Pickett, S.T.A. and J. M. Grove. 2009. Urban ecosystems: What would Tansley do? Urban Ecosystems 12:1-8.
34. Platt, R. 2006. The Humane Metropolis. University of Massachusetts Press, Amherst, MA. 368 pp.
35. City of New York. 2007. PlaNYC: A Greener, Greater, New York, Mayor Michael R. Bloomberg, The City of New York. http://www.nyc.gov/html/planyc2030/html/home/home.shtml (Accessed 05/31/2010).
36. Pouyat, R.V., J. Russell-Anelli, I.D. Yesilonis, and P.M. Groffman. 2003. Soil carbon in urban forest ecosystems, pp. 347-362. In Kimble J.M., L.S. Heath, R.A. Birdsey, and R. Lal (Eds.). The Potential of U.S. Forest Soils to Sequester Carbon and Mitigate the Greenhouse Effect. CRC Press, Boca Raton, FL.
37. Pouyat, R.V., I. D. Yesilonis, J. Russell-Anelli, and N. K. Neerchal. 2007. Soil chemical and physical properties that differentiate urban land-use and cover types. Soil Science Society of America Journal 71:1010-1019.
38. Rapoport, E.H. 1993. The process of plant colonization in small settlements and large cities. In McDonnell, M.J. and S.T.A. Pickett (Eds.). Humans as Components of Ecosystems: the Ecology of Subtle Human Effects and Populated Areas, Springer-Verlag, New York, NY.
39. Rees, W. and M. Wackernagel. 1996. Urban ecological footprints: Why cities cannot be sustainable—And why they are a key to sustainability. Environmental Impact Assessment Review 16:223-248.
40. Rudnicky, J.L. and M.J. McDonnell. 1989. Forty-eight years of canopy change in a hardwood-hemlock forest in New York City. Bulletin of the Torrey Botanical Club 116:52–64
41. Schleuß, U., Q. Wu, and H.P. Blume. 1998. Variability of soils in urban and periurban areas in northern Germany. Catena 33:255–270.
42. Shannon, C.E. 1948. A mathematical theory of communication. Bell System Technical Journal 27:379-423, 623-656.

43. Short, J.R., D.S. Fanning, J.E. Foss, and J.C. Patterson. 1986. Soils of the Mall in Washington, DC: I. Statistical summary of properties. Soil Science Society of America Journal 50:699–705.
44. Turner, W.R., T. Nakamura, and M. Dinetti. 2004. Global urbanization and the separation of humans from nature. Bioscience 54:585-590.
45. U.S. Census Bureau. 2000. Census 2000. http://www.census.gov/main/www/cen2000.html . (Accessed 05/31/2010)
46. U.S.D.A., N.R.C.S. 2010. The PLANTS Database. National Plant Data Center, Baton Rouge, LA. http://plants.usda.gov (accessed 06/08/2010).
47. U.S. Forest Service. 2007. Forest Inventory and Analysis National Core Field Guide. http://fia.fs.fed.us/library/field-guides-methods-proc/docs/core_ver_4-0_10_2007_p2.pdf (accessed 10/10/2010).
48. Uva, R., J. Neal and J. Ditomaso. 1997. Weeds of the Northeast. Cornell University Press, Ithaca, NY.
49. Whyte, William H. 1980. The Social Life of Small Urban Spaces. Conservation Foundation, Washington, D.C.
50. -----. 1988. City: Rediscovering the Center. Doubleday, New York

Appendices available online at http://digitalcommons.lmu.edu/cate/vol3/iss1/7/

AUTHOR NOTES

CHAPTER 1

Acknowledgments

We thank Coloma Rull, Margarita Parès, Montserrat Rivero, and Teresa Franquesa from the Department of the Environment of the Barcelona City Council for their support in this research. We also thank our colleagues of CREAF, especially José Ángel Burriel, for their support in GIS methods and mapping. Further our thanks go to i-Tree tools team (www.itreetools.org), especially to Al Zelaya, from the Davey Tree Expert Company, for their technical assistance with i-Tree Eco model. Finally, we thank the reviewers from the URBES project. This research was partially funded by the ERA-Net BiodivERsA through the Spanish Ministry of Economy and Competitiveness project "URBES" (code PRI-PIMBDV-2011-1179), by the 7th Framework Program of the European Commission project "OpenNESS" (code 308428) and by the Barcelona City Council.

CHAPTER 2

Acknowledgments

Funding was provided by EPSRC Sustainable Urban Environments grants EP/F007604/1 and EP/I002154/1. Infoterra kindly provided LandBase, and MasterMap was supplied by the Ordnance Survey. Leicester City Council provided GIS data delineating council land across the city, and J.J. Potter provided technical assistance.

CHAPTER 3

Acknowledgments

This work was funded in part by a TKF Foundation grant to MGB, an internal grant from the University of Chicago to MGB and the Tanenbaum

Endowed Chair in Population Neuroscience at the University of Toronto to TP. Data used for this research were made available by the Ontario Health Study (OHS), which is funded by the Ontario Institute for Cancer Research, the Canadian Partnership Against Cancer, Cancer Care Ontario, and Public Health Ontario. We thank the participants in the Ontario Health Study. We also thank Kelly McDonald and Tharsiya Nagulesapillai for preparing the data from OHS, and Ruthanne Henry for helping us gain access to the Toronto GIS data.

CHAPTER 4

Acknowledgments

We would like to thank the City of New York's Department of Parks and Recreation's Forestry, Horticulture, and Natural Resources Division for their support, and Fiona S. Watt, Jennifer Greenfeld, Jackie Lu and especially Anne Arrowsmith for their contributions to this research. We would also like to acknowledge the voices of those who spoke to us through their correspondence, without who this research would not have been possible.

CHAPTER 5

Acknowledgments

We gratefully acknowledge the support of the National Science Foundation Long-Term Ecological Research Program (DEB-1027188) as well as the assistance of the Parks and People Foundation and the City of Baltimore, Division of Forestry. We would also like to thank the three anonymous reviewers for their constructive comments.

CHAPTER 6

Acknowledgments

The authors thank Sarah K. Mincey and Matt Patterson, co-creators with the authors of the Protocol who participated in the deliberation over and design of variables. We are exceptionally grateful to current and former staff at Keep Indianapolis Beautiful who were supportive of Protocol development and testing (David Forsell, Jerome Delbridge, Nate Faris, Andrew Hart, Bob Neary, and Molly Wilson), and to the members of the KIB Youth Tree Team who tested the Protocol data collection methods and kept excellent field notes. We are also grateful to current and former members of the Bloomington Urban Forestry Research Group and Center for the Study of Institutions, Population, and Environmental Change, who contributed to refinement of Version 1.0: Rachael Bergmann, Kaitlyn Mc-Clain , Nick Myers, Shannon Lea Watkins, and Sarah Widney. Development of the Protocol and pilot research was provided by the Efroymson Family Fund, Keep Indianapolis Beautiful, Inc., the City of Bloomington Parks and Recreation Department, the State of Indiana Division of Forestry Community and Urban Forestry Program, and the Garden Club of America Zone IV Fellowship in Urban Forestry. Funding for printing of Version 1.0 was provided by the Alliance for Community Trees.

CHAPTER 7

Acknowledgments

DFD-P was supported by scholarships from the SEP-PROMEP and CONACYT, Mexico. KLE and KJG were funded by a Biodiversity and Ecosystem Service Sustainability consortium grant awarded through NERC (NE/J015369/1 and NE/J015237/1).

Conflicts of Interest

None declared.

CHAPTER 8

Acknowledgments

The research presented in this paper was funded by the National Urban & Community Forestry Advisory Council and the TREE Fund. The authors would like to thank the many people that made this project possible: Fiona Watt and Ayla Zeimer, New York City Department of Parks & Recreation; Jason Grabosky and Jessica Sanders, Rutgers University; Brian McGrath, Parsons The New School for Design; all the interns who helped collect data for this project.

CHAPTER 9

Author Contributions

MA analyzed the data. MA and DP wrote the manuscript. DP, SP, GJ, and TG designed the experiment. HM and LC collected the data. All authors edited and revised drafts of the manuscript, approved the final version and agree to be held accountable for the work.

Conflicts of Interests

The authors declare that the research was conducted in the absence of any commercial or financial relationships that could be construed as a potential conflict of interest.

Acknowledgments

This research was supported by National Science Foundation grants DEB 0919381 and 0919006, IOS 1147057, and EAR 1204442.

CHAPTER 10

Acknowledgments

The authors thank Mark Bradford, Bram Gunther, Minona Heaviland, Lea Johnson, Steward T.A. Pickett, Richard Pouyat, Kirsten Schwartz, Fiona Watt, Thomas Whitlow, Ian Yesilonis, and three anonymous reviewers. We also recognize generous support from the NYC Department of Parks & Recreation, The New School Tishman Environment and Design Center, Christian A. Johnson Endeavor Foundation, Andrew W. Mellon Foundation, Yale University School of Forestry and Environmental Studies Hixon Center, and AECOM Design.

INDEX

C

D

E

For Product Safety Concerns and Information please contact our EU
representative GPSR@taylorandfrancis.com
Taylor & Francis Verlag GmbH, Kaufingerstraße 24, 80331 München, Germany

www.ingramcontent.com/pod-product-compliance
Ingram Content Group UK Ltd.
Pitfield, Milton Keynes, MK11 3LW, UK
UKHW022159190926
475702UK00003B/104

* 9 7 8 1 7 7 1 8 8 4 2 5 9 *